Francis Edmund Anstie

Stimulants and Narcotics, Their Mutual Relations

With special researches on the action of alcohol, aether, and chloroform on the

vital organism

Francis Edmund Anstie

Stimulants and Narcotics, Their Mutual Relations
With special researches on the action of alcohol, aether, and chloroform on the vital organism

ISBN/EAN: 9783744795432

Printed in Europe, USA, Canada, Australia, Japan

Cover: Foto ©berggeist007 / pixelio.de

More available books at **www.hansebooks.com**

STIMULANTS AND NARCOTICS,

THEIR MUTUAL RELATIONS:

WITH

SPECIAL RESEARCHES ON THE ACTION

OF

ALCOHOL, ÆTHER, AND CHLOROFORM ON THE VITAL ORGANISM.

BY

FRANCIS E. ANSTIE, M.D., M.R.C.P.

Assistant Physician to Westminster Hospital, Lecturer on Materia Medica and Therapeutics to the School, and formerly Lecturer on Toxicology.

PHILADELPHIA:

LINDSAY AND BLAKISTON.

1865.

PREFACE.

THE purpose of this work is, of necessity, imperfectly represented by its title-page. No sentence of moderate length would have expressed the whole scope of the inquiry; a few prefatory words are therefore particularly needed, in order that the reader may appreciate the objects aimed at.

These objects are two—to destroy and to reconstruct; or rather to show, in some degree, how reconstruction may be possible in the future.

The historical chapter on the origin of the doctrine of stimulus makes no attempt at an exhaustive treatment of its subject, my only aim being to prove clearly that the source of the Vitalistic notions on which our classification of remedies is based is to be found in certain metaphysical theories. The dynamical terms which many modern writers employ ought not to be permitted to deceive us; they are, in fact, but the artificial clothing of a mode of research which is unconsciously but purely metaphysical, and, as such, extremely unfitted for the purposes of the physiologist or the physician. It is the mixture of a phraseology based on this kind of speculation with a perfectly skeptical empiricism in practice, which has brought the science of therapeutics into the extraordinary state of confusion in which we see it at the present day. It is the profound conviction that a complete review of our principles of classification is necessary, if experimental inquiry is to bring us any advantage, which has induced me to write the first two hundred and forty pages of the present volume.

The special researches which make up the latter part of the work are intended as the first of a series of *pièces justificatives*, which I hope to publish as evidence that the views promulgated in this book are not without solid foundation in observed fact. They by no means include all my researches on the properties of alcohol, chloroform, and æther, but only such of them as bear more directly on the questions treated in the other parts of the volume. If fault be found with their arrangement or the manner of their relation, it must be remembered that the difficulty was great of selecting, from a large quantity of material which I had collected, such portions as would best illustrate my subject within the compass which a work like the present affords.

For the special use of the word Hypnotism, which I have employed as signifying the production of natural sleep, and not in its modern and very inaccurate application to the phenomena of mesmerism, I think I need hardly make any apology.

On another matter I must say a word. In one or two instances, owing to the progress of science during the writing of this volume, an author is quoted only partially in the early sheets, and more fully in the later ones (this was unavoidably the case with regard to the opinions of Dr. Radcliffe). A glance, however, at the table of contents will enable the reader to follow with so much ease the general course of the work, that I believe no serious misapprehension will arise from it.

One feature in the work is sufficiently important to justify my calling the reader's attention to it; namely, the presence, among the notes at the end of the volume, of two bibliographical indexes, which may be of considerable use, I believe, to any one who desires to follow up the line of thought developed in my book.

To my numerous friends who have helped me, my warmest acknowledgments are due. To Dr. Reynolds the book may be said to owe its birth, since his kindness enabled me to solve the most serious preliminary difficulties as to its form and publication. To Dr. Radcliffe's writings I am indebted (in a degree which the pages of my book will testify) for materials of thought and experiment, and to his kindness for many additional hints, and one or two valuable cases. Dr. Buzzard, my colleague Mr. H. Power, Dr. Way, J. C. Whitehorne, Esq., and other valued friends, have wearied themselves in advising and in correcting ; and last, not least, I have to thank H. Chandler, Esq., of Pembroke College, Oxford, Dr. Meryon, and the Rev. H. Wace, for suggestions as to the sources of the necessary information, on many historical points, which were invaluable. For everything in the book, however, I am solely responsible.

16 Wimpole Street, Cavendish Square.
March. 1864.

TABLE OF CONTENTS.

INTRODUCTION.

CHAPTER I.

HISTORY OF THE DOCTRINE OF STIMULUS AND ITS PHILOSOPHICAL ORIGIN IN THE VITAL THEORIES OF THE ANCIENTS.

CHAPTER II.

CRITICISM OF THE DOCTRINE OF STIMULUS.

This doctrine assumes that—

  All mental excitement;

  All increased sensibility, and pain;

  All convulsive muscular action;

  All considerable increase of secretion;

  All increase, whether of force or frequency of the heart's action,

Are caused by, and the proofs of, a stimulant action upon the organism   .   83

CHAPTER III.

SUGGESTIONS FOR THE RECONSTRUCTION OF THE DOCTRINE OF STIMULUS.

CHAPTER IV.

THE DEFINITION OF NARCOSIS.

CHAPTER V.

SYMPTOMS OF NARCOSIS.

CHAPTER VL

ON CERTAIN BODILY CONDITIONS WHICH ARE UNFAVORABLE TO THE PRODUCTION OF NARCOSIS.

CHAPTER VII.

ON THE RELATION WHICH STIMULATION AND NARCOSIS BEAR TO EACH OTHER, IN THE ACTION OF THOSE SUBSTANCES WHICH ARE CAPABLE OF PRODUCING BOTH 221–224

CHAPTER VIII.

CHAPTER IX.

GENERAL CONCLUSIONS.

SPECIAL RESEARCHES.

PART I.

ÆTHER-NARCOSIS.

PART II.

STIMULANT ACTION OF ÆTHER.

2

RESEARCHES ON THE ACTION OF CHLOROFORM.

PART I.

THE PHENOMENA OF NON-FATAL CHLOROFORM-NARCOSIS.

PART II.

PHENOMENA OF FATAL CHLOROFORM-NARCOSIS.

PART III.

THE STIMULANT ACTION OF CHLOROFORM.

RESEARCHES ON THE ACTION OF ALCOHOL.

PART I.

NARCOTIC EFFECTS OF ALCOHOL.

PART II.

TRUE ALCOHOLIC STIMULATION.

NOTES.

STIMULANTS AND NARCOTICS.

INTRODUCTION.

AMONG the too frequent instances which are to be found in medical nomenclature, of confusion and uncertainty in the application of descriptive terms, there is none, perhaps, more striking than the position which is occupied by the words "Narcotic" and "Stimulant," in our theories of the action of remedies upon the body. To the layman, indeed, these words have a definite meaning, for they represent, respectively, certain stupefying poisons, and certain grateful restoratives. To the busy practitioner, again, who has no time to theorize, but who must cure his patients, the former word stands for a number of medicines which will relieve pain and procure sleep, and the latter for a class of remedies which he has frequent need to use when he desires to produce a rapid revival of vital powers which are temporarily depressed. But to the philosophic student of medicine, who desires to arrange in orderly classification the weapons of his art, and thereby to multiply its resources, the accurate definition of these two classes of remedies offers a problem at once of great interest and of extreme difficulty; for, although the *typical* features of Narcotism may be easily exemplified by the results of such an agent as chloroform, administered till it produces unconsciousness, and those of Stimulation, by the effect of a glass of wine in rousing a man from a fainting fit, we are not able from these examples to conclude, simply, that chloroform is a narcotic, and that wine is a stimulant. More extended observations would teach us that, under certain conditions, these

3

effects might be reversed, and that chloroform might prove a reviving stimulus, and wine a stupefying poison, capable of rapidly destroying life. Renewed inquiries would at last elicit the fact that an intermediate class of medicines exists, the members of which somehow possess both stimulant and narcotic properties, and that to this class both chloroform and alcohol belong. At this degree of information medical science long ago arrived; and beyond it, in the way of intelligibly explaining these curious facts, it does not appear to have passed. Thirteen years ago, Mr. Simon, in the course of a series of lectures on Pathology, made the following remarks, which I confess appear to me equally applicable to the present state of things: "Our therapeutic control over the excitement of nervous centers is hitherto imperfect, less for want of powerful agents than because of our want of precise knowledge as to their distinctive effects. The so-called sedatives, acting through the blood on the higher nervous centers, almost always commence their operation by stimulating those centers; and the so-called stimulants almost always produce a subsequent sedative effect; so that it is not easy to determine whether the two classes of medicine can be separated." These words still correctly depict the obscurity which distinguishes all our ideas upon this most important question.

The following considerations have induced me to attempt a reinvestigation of this subject. Although no sufficient clew appears to have been given, by any of the eminent writers on Therapeutics who have rendered such important services to practical medicine, for the elucidation of these matters, yet there are many facts, gleaned by their diligence and acuteness of observation, which seem to suggest important inferences as yet overlooked. Reflecting on these facts, I have been induced to make some fresh inquiries, the results of which, I think, are not without value, and which appear to confirm the inferences which I had already been inclined to draw from previously recorded observations.

Moreover, it has appeared to me, in studying the various formal essays upon therapeutics which have appeared of late years, that a certain timidity has, in many instances, induced their authors to cling to the use of phrases which represent physiological theories now virtually dead; and that the use of these phrases has injuriously affected the value of our systematic teaching on the subject of the action of medicine, leading energetic practical men to regard it as somewhat pedantic and unreal. I cannot but think, that by bringing the light of recent clinical observation and physiological experiment to bear on our therapeutic classifications, we may effect some real improvement in the latter. Desiring to speak with all possible respect of the truly great and valuable works of such men as Thompson, Christison, Pereira, Headland, and others, I believe it is possible, by degrees, to improve considerably on the results obtained by them; more especially in the direction of the simplification of ideas.

In order that we may start fairly on the inquiry which is before us, let us understand what are the acknowledged elements of the problem to be solved. They are simply—the existence of two classes of physiological agents, respectively known as Stimulants and Narcotics (or Sedatives), and of an intermediate class, known as Narcotic-Stimulants (or, in looser phraseology, as Narcotics); all three classes acting upon the nervous system: the stimulants having the power of exciting its action; the sedatives, of depressing the same; and the narcotics, or narcotic stimulants, of producing both kinds of effect. It is to the latter class, as forming the meeting-point of the two kinds of physiological action, that our closest attention will be required: and it is to this class that the three substances which have been chosen for detailed investigation belong; as do also tea, coffee, tobacco, and the whole genus of soothing, "care-breaking" luxuries (to use an expression of Von Bibra's), so freely used in every-day life. All these classes act upon the nervous system, either directly, as when the nerves affected lie close beneath the skin or mucous

membrane to which the agent is applied, or indirectly, as is
more usually the case, by being absorbed into the blood,
and carried by the circulation to all parts of the nervous
apparatus.

So much is agreed. What remains doubtful is not so easily
expressed; but we may say this, that the difficulty would be
pretty nearly solved, if once we could come to an understanding as to what effects may fairly be included under the general term Stimulation, and which may not be so included.
The remaining phenomena would then fairly settle down into
their places, and the substance which produced them might
be ranked either in one or two pretty well-defined groups.

The present chapter will be occupied with some preliminary
observations on the tone of therapeutical inquiry as it exists
among us, which must be made before entering into the
discussion of any particular theories about narcotics and
stimulants.

It may be said, in general terms, that this kind of inquiry
is, and always has been, more affected by the injurious influence of traditions, handed down from a source which is far
removed in the gloom of antiquity, than any other branch of
medical investigation. It is difficult for any one to imagine,
who has not studied the various formal treatises on Therapeutics in something like an historical order, with what deference we all receive statements made in the axiomatic form,
simply because the ancients thought that they required no
special proof. From the following instances, the correctness
of this statement will be perceived. The first is the fact—
that it is the general custom with writers on Therapeutics to
assume that the same qualities which distinguish the action
of any drug when administered in a large dose, must distinguish it when administered in a smaller quantity, and *vice
versâ;* the only difference being one of degree, and not of kind.
Much of what I shall have to say on this point is not new,
having been already stated by others, or by myself, in other

places ;* but as the proposition in question is just one of those vagabond assertions which are apt to elude all sentinels, and slip unperceived into camp, I must, for the sake of completeness, include a refutation of it in my **argument.**

The source of the fallacy is not far to seek. Certain common medicinal remedies are capable, when administered in larger than medicinal doses, of producing poisonous effects, which *appear* to be only the exaggeration of the ordinary medicinal action. For instance : Opium produces in a medicinal dose, sleep ; in a poisonous dose, coma. The latter state, it is argued, is only the extreme development of the former ; *ergo*, the action of opium is *essentially* the same, in all doses. Again, it is said, most of the vegetable aperients are capable of causing, when given in very large doses, such extreme purging, and other results of intestinal irritation, as to prove fatal to life : this action is evidently continuous with the ordinary medicinal effects **of these drugs.** It is hardly necessary to point out to a physiologist the wide gaps that exist in the reasoning of those who argue thus, or to remind him **that** there is more than a doubt whether coma be exaggerated sleep **at** all, and whether death from vegetable " irritants " is due solely to excess of irritant action on the alimentary canal.

But it is not on mere negative arguments that the question turns. If the dogma, that the action of all doses of a drug is the same essentially, though not in degree, be true in any such sense as would warrant its being assumed as a basis of therapeutical inquiry, it ought to be true universally. So far from this being the case, it is possible easily to prove that it is false in half a dozen conspicuous instances.

Common salt is, in small doses, a perfectly indispensable article of human food, without which we should perish miserably ; in medium doses, it is a safe and useful emetic medicine ; while in extremely **large** doses, it is an irritant poison, **and has** caused death† **in several cases. Now, it** is plain

* " London Med. Review," 1862. " Cornhill Magazine," June and September, 1862.

† *Vide* " Med. Gazette," 1839–10, vol. i, p. 559. Christison "On Poisons," 658.

that the first and second of these actions, at any rate, are not physiologically continuous. The food-action of salt consists of a series of chemical and vital processes, by which that substance, being absorbed into the blood, in part remains as one of the normal elements of the blood, and partly is decomposed into acid for the gastric juice, and soda for the bile. The irritant and inflammatory effects produced by larger doses are clearly quite distinct from this, and belong to a separate series of vital processes; or rather, they are not vital at all, but morbid.

Again: Iron, in small quantities, is quite as important an element of our daily food, and is indispensable to health as is common salt; for it is a normal element of the blood, and a deficiency of it is a frequent cause of the state which is called anæmia. But iron, in very large doses, such as two or three ounces of green vitriol (the proto-sulphate), or of the tincture of the sesquichloride, is capable of causing fatal inflammation of the stomach and bowels.* There is clearly no analogy whatever between the action of iron as a food, or a food-medicine, and its action as a poison: the two things are radically distinct.

The various effects of arsenic afford another refutation of the dogma we are considering. Speaking merely from the results of medical practice, in England, we can safely say that, in doses of from one-thirtieth to one-twentieth of a grain, this substance may be taken two or three times a day, for months together, without producing any damage to the system; indeed, with the most beneficial results, in proper cases, in improving the quality of the blood and the general condition of the body. On the other hand, every one knows that arsenic, in large doses, is a powerful irritant poison, producing a specific inflammation of the stomach and intestines, which proves rapidly fatal.

---

* Vide Christison "On Poisons," 506. Orfila, "Ann. d'Hygiène," 1851, vol. ii, p. 337. "Med. Gazette," vol. xlvii, p. 307. "Ann. d'Hygiène," 1850, vol. i, pp. 180, 416. Taylor "On Poisons," p. 559, &c., &c.

But there is no need to multiply examples; they are to be found on every side. Indeed, it is hardly too much to say that we cannot state with certainty of any food, that it may not also be a medicine and a poison; nor of any poison, that it may not also be a medicine and a food, under some circumstances. The celebrated maxim of Boerhaave, that nothing can be properly called a remedy "quin solo tempestivo usu," is, in fact, only part of this wider truth, which is, however, too often lost sight of. Fortunately, the practice of the medical art is in advance of the theories which are still propounded in books; and, while the formal teaching of the day, for the most part, separates physiological agents into rigidly-defined groups, as foods, medicines, and poisons, the practical tact of medical men is rapidly discovering, that the propriety of applying these terms depends entirely upon the conditions under which any particular substance is administered, and that among these conditions *dosage* holds the highest rank.

It is certainly not a general law, then, that the essential character of the effects produced by physiological agents remains the same, although their degree may be altered, whatever doses may be employed. We shall have occasion to remark, hereafter, that the assumption that this was the case has exercised a strong influence upon the prevailing ideas as to the action of those drugs which possess both stimulant and depressing powers.

The second objection which, I think, fairly lies against nearly all existing descriptions of narcotics and stimulants, and of their mutual relations, is, that they assume that the essential characters of the two kinds of physiological action, respectively, are to be found in the presence of certain striking symptoms. Consciously, or unconsciously, nearly every writer gauges the stimulant power of drugs by their ability to quicken the pulse and raise the temperature of the surface; and their narcotic influence, by their power to relieve pain and to induce sleep, or to reduce the frequency of the circulation. It will be necessary for me to defer to a more advanced point in my

argument the full examination of the fallacy which, in my
opinion, lies at the root of these ideas; at present, I wish to
call attention to the fact, that this manner of reasoning upon
the action of drugs is based on the assumption, that all
striking vital phenomena—everything by which the bodily
nature asserts itself strongly, and, as it were, obtrusively,
indicates exalted functions in an organ, or exalted force in
the organism generally.

Upon such assumptions was based the theory of true
"sthenic" inflammation as a bodily condition evidenced by
four distinctive symptoms—heat, swelling, redness, and pain:
the joint presence of all these phenomena being supposed to
give assurance that the affection was of the sthenic (or strong)
type. It requires a considerable effort of imagination to appreciate the distance at which we stand, in the present day,
from the pathology of the times when these laws of diagnosis
were propounded; for through the influence of a spirit of
conservatism, which it is difficult to know whether to praise
or blame, these definitions of sthenic inflammation still hold
their place in our text-books, when their original significance
has entirely disappeared. I confess I think this is a great
misfortune. One evil consequence of it is, the perpetuation
of a mode of thought which looks upon vital force as something separate from, and independent of, the organs which
manifest it, and liable to capricious changes in its development which call for a good deal of interference, by extraordinary means, to restore the true balance of things. Such
a disposition of mind must surely be a very bad preparation
for inquiries in therapeutics. It must certainly tend to make
the physician overlook the important question as to how much
of any given effect is due to his remedies, and how much to
nature, and to direct his attention from that reverent and
patient observation of natural processes without which no
medical man ever did great things for mankind, or for the
advance of his art.

I would ask the candid reader to consider whether we are

not still unconsciously acting under the influence of some such notion as this—that there is one degree of vital action, which may properly be called Healthy Excitement, a second, somewhat higher, which is Irritation, and a third, **the highest** of all, which represents true Inflammation? Do we not feel impelled, by the force of imagination, to interpret the signs of acute disease as signs of vital energy, the fact being that **we are thinking, not** of degrees of action as they really occur in the body, but of degrees in an imaginary scale which **we** erect in our own minds?

These sources of fallacy have been noted by Dr. Chambers in his recent work on the "Renewal of Life," a work which requires some notice here, because it contains, along with much which I cannot agree with, certain views of disease which point to a comparatively new and unworked field of therapeutical inquiry.

The position which Dr. Chambers seeks to establish **is** this: "That disease is, in all cases, not a *positive existence*, but a *negation;* not a new *excess of action*, but a DEFICIENCY; not *a manifestation of life*, but PARTIAL DEATH; and therefore that the BUSINESS OF THE PHYSICIAN is, directly or indirectly, not to *take away* material, but to ADD; not to *diminish function*, but to GIVE IT PLAY; not to *weaken life*, but to RENEW LIFE."* I have preserved the original typography, in order to give all the emphasis to this doctrine which its author could desire.

It happens that my own attention has for a long time past been directed to the *alimentary* character of numerous drugs, and to the fact that modern practice includes a constantly increasing selection of such remedies in its pharmacopœia. Of the truly alimentary character of some medicines, such as cod-liver oil and iron, there is no longer any dispute, though there are differences of opinion as to the mode in which their food-action takes effect. But with regard to many other substances, there is the greatest opposition of views, and especially

* "The Renewal of Life." By Thomas K. Chambers, M.D., &c. Churchill, London; Lindsay and Blakiston, Philadelphia.

with respect to the whole class of narcotic-stimulant medicines, and of their congeners—tea, coffee, tobacco, and the
like, which are in use as articles of every-day consumption.
I cannot but think that a work professing to treat of such
vast questions as those propounded by Dr. Chambers should
have included something more serious on this subject than
the short and rather superficial chapter on Alcohol at the end
of the volume.

It so happened, that owing to the circumstance of my being
for several years (between 1855 and 1861) in the constant
habit of administering chloroform, I became interested in the
action of anæsthetic agents of various kinds, and made a
great many experiments, both on animals and on the human
subject, and I particularly investigated the action of small
doses of these substances. Some of the results obtained in
the latter way were so unexpected that they led me to extend
my inquiries to various other narcotic agents, with particular
reference to the influence of alterations of dosage; and the
consequence has been that I have been led to propose a different explanation of the medicinal effects of many narcotics
in common use from that which is ordinarily stated.

I wish to call the reader's attention to the remarks and to
the clinical illustrations, which he will find in the section on
Stimulation, on the production of sleep and the arrest of convulsive movements by medicinal doses of opium, chloroform,
sulphuric æther, and other remedies which in common par-
lance are called narcotics, and to the comparison instituted
between the action of these substances, given in this way,
with that of undoubted stimulants, such as ammonia, and
with that of common foods. I feel that it is impossible for
me to do full justice to this very important subject, but I
should be only too glad to initiate further inquiry. It has
surprised me very much to find that more notice has not been
taken of this aspect of the question by authors who have
written expressly on the Use and Abuse of Narcotics; especially as the facts of the case, although in an isolated form,

have been noted by various writers, and only needed careful collation and comparison to suggest important **inferences**. Some of the remarks of Dr. John Brown (hereafter to **be** quoted) on this topic are pregnant with interest, although **he** unfortunately thought it necessary to incorporate them **with a** highly complex theory, which was destined **to go the way** of all such devices for explaining every difficulty in medicine, and **converting it into a science of certainties.**

The profession, and the public generally, **have a right to** complain **that the writers of systematic** treatises on **Thera**peutics have shown too great a disregard to certain broad and important facts as to the **daily use of the so-called nar**cotics (which, in truth, are also stimulants) by persons **in a** state of ordinary health; **and it is necessary to state these** facts clearly, **even at the risk of saying what is not new to** the reader, because they **have a direct bearing on the par**ticular views which will be put forward in the course **of this work.** The first **of these is**—that there is no **nation, and that** there are very few individuals, who do not **make daily use of** some substance **to which** the term Stimulant Narcotic may be applied in strictest accordance with what we know **of the action of drugs.** Von Bibra* puts the matter roughly, **but** strikingly:

" Coffee-leaves **are** taken, in the form **of infusion, by two** millions of the world's inhabitants.

" Paraguay tea is taken by ten millions.

" **Coca** by as many.

" **Chicory,** either pure or mixed with coffee, **by forty** millions.

" **Cacao, either** as chocolate, or in some other **form, by fifty** millions.

" Haschisch is eaten and smoked by 300 millions.

" Opium by 400 millions.

" Chinese tea is drunk by 500 millions.

" Finally, all the known nations of the world are addicted

* **Von Bibra,** "Narkotischen Genussmittel und der Mensch"—Preface.

to the use of tobacco, chiefly in the form of smoke ; otherwise, by snuffing or chewing."

Professor Johnstone* completes the picture thus drawn, by an ingenious map, in which it is sufficiently shown that no considerable tract of the earth's surface is without some special indigenous narcotic plant, of which the natives freely avail themselves, not merely for medicinal purposes, but for every-day use.

Nor is the use, in every-day life, of these substances an outgrowth of modern corruption ; on the contrary, it is consecrated by whatever sanction immemorial custom can confer. There is absolutely no period of history, as there is absolutely no nation upon earth, in which we do not find evidence of this custom.

Against these weighty statements there is nothing to be said ; they are as free from any admixture of fallacy as they could possibly be. It is idle to urge that the subject of a carefully prepared experiment can be made to live in apparent health without the use of any of the substances vulgarly called Narcotics, if the practical fact be that *nations* cannot, and never have been able to, do without them.

Under these circumstances it is the duty of the physiologist to inquire very seriously whether there is not some practical test by which we may distinguish the effect which is hurtful, from the effect which the instincts of men teach them to believe is useful. But this task is of immense difficulty. It can only be approached, with the remotest chance of success, by means of a simultaneous use of the facts which are known, or can be discovered, as to the action of the narcotic-stimulants in disease, of those which the every-day experience of a large number of healthy consumers of narcotic-stimulants supplies, and of those which transpire when the victims of the abuse of these powerful physiological agents come under the care of the physician. I need hardly say that it is necessary entirely to exclude the special moral and religious aspects

* "Chemistry of Common Life."

of the question if we wish to arrive at a sound, scientific result. What we desire is not a test of this kind—which in human hands is liable to be so much perverted—but something positive, tangible, capable of physical demonstration.

Such a test there must surely be, if we could only find it. It is inconsistent with our ideas of the order and design of the universe, that so important a matter as the discrimination between foods and poisons should be left to chance or instinct, without the possibility of certain knowledge. Our best hope of arriving at this knowledge lies in studying diligently the nature of those actions which we call stimulant, and of those which we call narcotic, respectively, in order that we may analyze that compound result which is observed to be produced by the administration of those substances which are capable of producing both these effects. I venture to think that, notwithstanding the frequency with which we use both these terms, we seldom realize to ourselves what we mean by them. I think so for this reason—that we constantly hear people designate the same physiological actions interchangeably by both these names; and that it is easy to trace, in the familiar expressions we use in common conversation or ordinary clinical description, that, in fact, we do not recognize the distinction between them. In this way the greatest confusion arises, both in our physiological and our therapeutic notions; and the consequence is that we seldom realize fully, when prescribing one of those remedies which stand on the debatable ground, whether we intend to produce a stimulant or a narcotic effect.

Another source of much obscurity in our ideas is the use of the word Sedative in its modern sense. The action of "Sedatives" has been made to signify something different from the depressing part of the influence exerted on the nervous system by narcotic-stimulants; it has been supposed that sedatives act chiefly upon the sensitive nerves, and that therefore they are to be distinguished from all other depressors of nervous force. How unreal this distinction is, I shall hereafter endeavor to show; meantime it has occasioned a great deal of trouble and anxiety to the student, who might otherwise have contented himself with using the word "stimulant" to express an agent which exalts nervous force, "narcotic" to signify one that depresses it, and narcotic-stimulant for one which produces both kinds of effect.

The plan which will be adopted in the ensuing inquiries is as follows: The first section will be devoted to the consideration of various theories of stimulation which have been proposed, and I shall endeavor to give an explicit idea of the actions which ought to be included under that general name, illustrating my remarks by clinical and physiological observations. The general features of sedative action will next be sketched, the common features which distinguish all varieties of this effect indicated, and the apparent contradictions as far as may be reconciled. The mode in which the two kinds of physiological effect are combined in the narcotic-stimulants will be then described; and a short chapter will be given to those substances which produce distinctly irritant poisonous effects, simultaneously with their narcotic effects.

As it would be clearly impossible, on the mere score of space, if for no other reason, for me to dwell minutely on the individual members of such very large classes of medicines as the stimulants and the narcotics, I shall content myself, in the body of the work, with isolated examples taken from the action of the more prominent members of these groups. The latter portion of the volume will be taken up with a more extended series of observations on the three principal members of that group of narcotic-stimulants commonly known as the "Anæsthetics," viz., alcohol, chloroform, and ether. These substances have been chosen for detailed examination, as well from their great importance as from the fact that many features of their action upon the organism are already familiar to most readers.

I feel that it is necessary for me to bespeak a very patient hearing. Some of the views which I shall have to bring forward are opposed to ideas which have become, by the influence of tradition, incorporated into the stock of things which are received as matters of course. I can only say that a considerable amount of experimental familiarity with the subjects of the present inquiry has convinced me that more is commonly taken for granted than will stand the test of proof in regard to these matters; and that this is particularly to be noted, that the ordinary mode of reasoning on the effect of physiological agents is vicious, *ab initio*, because it retains, in many cases, the unmistakable impress of metaphysical speculation applied to things with which it has really no fit connection. My object will be fully attained if a more general interest be awakened in the profession, in the bearing on therapeutics which the facts of common life, apart from the circumstances of actual disease, assuredly have. It is no stigma on the reputation of those who have already handled the subjects which I am now to deal with, that their writings exhibit, as I think they do, a too limited view of the scope of medical inquiry. It was but yesterday that disease was universally regarded as something entirely foreign to the vital organism, which came to it from without, resided in it for a time, and then departed, exorcised by the physician's art. To-day we are inclined to take a less exalted view of our functions, to confess ourselves to be but the humble assistants in those curative processes which Nature herself initiates, and very often carries through without our help, or even in spite of our ignorant interference. Together with such changed ideas there must come a revolution in our modes of therapeutical inquiry; and notably, a disposition to compare those instances of the beneficial action of drugs which are well authenticated, with similar effects produced by the unaided operation of natural causes. And it is surely lawful to hope that even partial success in this direction may prove more advantageous to the progress of our art than the most brilliant reasoning which should presuppose the physician's power to effect radical alterations in the working of the vital agencies, whose operations we are only just beginning dimly and partially to understand.

CHAPTER I.

THE history of the doctrine of Stimulus may be said to be the history of medicine itself. At least we find traces or foreshadowings of this doctrine in the discussions which agitated the medical schools of every age, if we except that remote period of antiquity in which, as yet, no theory of vital action, of the nature of disease, or of the *modus operandi* of remedies had appeared, and medical inquiry was confined to an empirical search for substances possessed of curative properties.

It need hardly be said, however, that the modern ideas of stimulation are widely different from the form in which the process first figured itself to speculative minds. So marked is the difference, that one* of the few writers who have attempted to trace the history of the doctrine refuses to allow the existence of any distinct theory on the subject prior to the sixteenth century, near the commencement of which certain authors spoke of the "irritation exercised by the humors on the solids." I believe this to be a partial and incorrect view. The truth seems to be, that the modern discussions on Stimulation, Irritation, Vital Excitation, and the like, are but a part of a much larger question, which has been debated in every age. "How is it that mind can enter into relation with matter, so that the two things, so utterly dissimilar, may react upon each other?" In the early speculations of great thinkers upon this endless topic, we may, perhaps, find the clew which we require.

The great father of the medical art—Hippocrates—lived in

---

* Broussais. De l'irritation et de la folie, p. 48. Paris, 1828.

a time and country in which such questions excited a vivid interest. In him were united the sacred dignity of the Esculapian priest; the high philosophic culture of the age and of the nation which produced his contemporaries, Socrates, Plato, and Aristotle; and a power of accurate observation of natural phenomena unrivaled in his own or any other time. That such a man would leave the mark of his thought upon the medical science of many succeeding centuries was inevitable: nor was it less certain that the thoughts thus handed down would reflect the philosophic spirit of his age, and that the example of their great teacher would incline his successors to study medicine, not as a thing apart, but as one branch merely of a general philosophy. It is, therefore, a matter of great interest to ascertain what were the views of vital action which were held by the great philosophers of his day, and how far they were modified by him in consequence of his practical acquaintance with vital and morbid phenomena.

The doctrine of Plato as to the relations of matter to intelligence in the human organism is expressed in the Timæus.* He there represents the Supreme Deity as committing into the hands of the inferior gods the intelligent soul, in order to the creation of man. These deities construct for it a mortal body, and form within this a kind of *mortal soul*, possessed of passions and appetites. This mortal soul is placed within the trunk of the body, and is thus separated, by the whole length of the neck, from the head, wherein is lodged the immortal part, in order that the divine nature may not be sullied more than is necessary by contact with the baser. Further, the mortal soul is again divided into two parts, whereof the nobler, which partakes of courage and spirit, and loves strife, is situated within the cavity of the chest: being thus more immediately under the control of the reason, which resides in the head; while the appetitive part is located in the abdomen (chiefly in the liver). The heart is placed in a

---

* Timæus, § 69–70. Ed. Stallbaum. Etudes sur le Timée de Platon. Par T. H. Martin. Paris, 1841. Notes 139, 167, 186.

4

kind of watch-house, and being connected with the members by means of the veins, it is able, when informed by the reason of any evil committed in the members (from any foreign cause, or any internal passion), to send through all its channels the threatenings and exhortations of reason, and thus reduce the body to perfect obedience. The lungs are placed around the heart, in order that being soft, and *bloodless*,* and spongy, they may cool the heart when it is inflamed with **passion**. The office of the πνεῦμα, or breath, is to cool the **blood**; and the various obstructions and diversions of its course to which it is liable are the cause of many diseases.

According to Aristotle, the vital principle of every living animal and plant resides in the germ.† He describes it as something which may be called heat; but which is not *fire*, but a spirit, which is of the nature of the sun and stars. For fire begets nothing, nor is anything composed by the dense,‡ the moist, or the dry; but the sun's heat, and the heat of animals (not only that which is in the germ, but also if there be any kind of excretion), this also has a vital principle.§ He says that the soul contains the body, rather than the body the soul, a doctrine which, as Sir W. Hamilton‖ remarks, was the basis of the common dogma of the schools, **that the** soul is all in the whole body, and all in every of its parts. This residence of the soul in every part of the body may be the reason, according to Aristotle, why severed portions of some insects may continue to show signs of vitality, **the** ψυχή itself being divisible. Of the functions of the brain as an organ of intellect, Aristotle knew nothing. He places the *mortal soul* in the heart, as did also Zeno, and the Stoics generally, at a later period.¶ Like Plato, he believed that

* This part of the description is, of course, exactly the reverse of the truth.

† **De Generatione Animalium, lib. ii, cap. 3.** De partibus Anim., lib. ii, cap. 2.

‡ **The** reader will recognize the reference to the Pythagorean doctrine of the four elements, which **was** held by the great majority of ancient philosophers.

§ De Animâ, lib. i, cap. 5.

‖ Notes on Reid, p. 856, foot-note.

¶ For an abstract of the Aristotelian doctrine, see Sprengel, Geschichte der Arznei-kunde, Th. i, p. 528–30.

one chief office of the πνεῦμα, or breath, was to cool the blood, and that it acted in some way as an important instrument of mind in its operations on the body.

The existence of a materialistic school of philosophy (as represented in Hippocrates' time by Democritus particularly) requires a word of notice here, in order that I may state that, so far as I can perceive, the Corpuscular theory had no real share in the foundation of a doctrine of vital irritation, either at this time or in its later development by the Epicurean philosophers, and by the methodic school of medicine founded by Asclepiades and Themison.   From first to last, in the history of medicine, the Corpuscular doctrines, which were in total opposition to the Hippocratic, were more or less overborne by the latter, which ultimately obtained a decided ascendancy, chiefly through the powerful advocacy of Galen.

Such was the philosophic atmosphere in which Hippocrates lived; its influence is very conspicuous in his writings.   He accepts the Pythagorean doctrine of the four elements (probably in a far more literal sense than did Plato or Aristotle), and he erects upon it the theory of the four principal humors of the body.   He recognizes the existence of an *intermediate* nature, the source of vital activities (τὸ ἐνορμῶν—the *impetum faciens* of later writers) which is distinct from the immortal soul.   He seems to identify the pneuma, or breath, with the very *soul* itself, at least such is the apparent meaning of a passage in his treatise on Epilepsy.*   He recognizes a *vis medicatrix naturæ*, in virtue of which the organism reacts against hostile influences, and by means of the processes which we call disease, tends to cure its own disorders.

In the first century of the Christian era, Athenæus of Attila founded a system of pathology, in which great prominence was given to the influence of the pneuma, and the supporters of which were thence called Pneumatists.   According to this school, the phenomenon of the pulse in arteries was due to

---

* De Morbo Sacro, xvii. Ab. Kaau Boerhaave. "Impetum faciens." Lugdun. 1745.

the alternate contraction and dilatation of those vessels during the passage of the pneuma. Aretæus, who at the commencement of his career was a pneumatist, says that the pneuma passes from the heart to the arteries, and by these channels to the whole of the body. It was universally believed, at this time, that the arteries did not contain blood, but only some kind of spirit; an error for which Praxagoras seems to have been primarily responsible.

The influence of Stoic teaching* caused the still more decided identification of the pneuma with the soul itself: as is very apparent in the account given by Galen (circ. A.D. 130) of the nature of the vital forces. He speaks of a natural force which resides in the liver, and is due to the influence of the πνεῦμα φυσικόν, a life-force which resides in the heart, and is nourished by the πνεῦμα ζωτικόν: and an animal-force (or soul-force), which resides in the brain, and is supplied by the πνεῦμα Ψυχικόν.† All these varieties of the pneuma are only modifications of the inspired air: the latter enters the brain by the nostrils (passing through the sieve-like plate of the æthmoid bone), it enters the heart with the blood which comes from the lungs, and it enters the liver by means of anastomoses which connect the vessels going to that viscus with the arterial system. The learned author of an article‡ in the *Allgemeine Encyklopädie* remarks, that we have in this doctrine of Galen, the germ of modern ideas as to the distinction between Vegetative, Irritable, and Sensitive life.

The influence of Galen was most powerful; it carried all before it, and for centuries was everywhere predominant in medical Europe. The decline of anatomical studies which followed his death, together with the influence of Alexandrian Neo-Platonism, tended to the production of many strange and

---

* Chrysippus—quoted by Galen. De Plac. Hipp. et Plat. ii, p. 264; iii, p. 1-7; vi, p. 2-9. Diog. Laert. lib. vii, c. 1, sect. 84, § 167. Plutarch. Adv. Stoicos, cap. 46. De Prim. frig. c. 2.

† Galen. De loc. affect. v. De usu part. lib. v. Art. et ven. diss. lib. vi, vii. De usu Respir. p. 163-4.

‡ Gruber's Allgemeine Encyklopädie. Art. Galenus.

fanciful developments of the pneumatic doctrine, as to which
a very interesting chapter might be written had we unlimited
time and space at our disposal.*   We must not, however,
pass without notice the doctrine of the celebrated Nemesius,
Bishop of Emesa in Syria (circ. A.D. 360). In his treatise,
"De Naturâ Hominis," we find it stated that the dilatation
of the artery "draws with force the thinner parts of the blood
from the next veins, the exhalation or vapor of which is made
the aliment for the *vital spirit*."†

At the downfall of the Alexandrian school of medicine
(641), caused by the taking of the city by the Saracens, and
the destruction of the great library, Greek learning and
Greek medicine might be thought to have received a death-
blow: on the contrary, however, the Saracens soon turned to
the Greek authors, and studied them with an almost slavish
reverence.   Galen was especially held in honor, and pneumatism seems to have been a universal creed among them.   It
is useless, however, to delay over this period of our history,
for the Saracenic epoch is a blank as regards any real progress
in physiology or pathology.   But it was favorable to the progress of pneumatism; and so, either directly or indirectly, was
another great influence which waged war with the Saracenic
power in Europe, and at last overthrew it—that of the
Christian Church.   Ecclesiastics by degrees acquired a great
deal of influence on medical education, and at last monopolized
it almost entirely; and it would seem that, whether Neo-
Aristotelian or Neo-Platonist, they almost universally favored
the belief in forces analogous to the pneuma.   It need hardly
be said that both Aristotelian and Platonic doctrines were
extremely corrupted, owing to the wretched translations, at

---

* The reader who may feel interested in this subject is recommended to study Vol.
II of Cudworth's Intellectual System: Plotinus, Enneades (A French edition by M.
Bouillot—Paris, 1857—has most valuable *notes*); Proclus Institutiones; John Philoponus in Aristot. De Animâ (there is a good Latin translation, published at Venice,
1600), Preface; Etudes sur le Timée de Platon, par T. H. Martin. But the subject is
endless, or rather, one might say, *bottomless*.

† Nemesius, De Naturâ Hominis. Ed. Matthiæ.

third hand, which were the only ones existing, in Western
Europe at least, and which were the guides even of such men
as St. Thomas Aquinas, and his master Albertus Magnus.
This corruption of the classic writers was doubtless partly
responsible for the mystic and magical notions which vulgar-
ized medical science, and tended to keep it in the hands of
privileged quacks.

At the time of the general revival of classical literature,
which took place under the auspices of the Greek literati
who crowded to the Medicean Courts in the fifteenth century,
both Plato and Aristotle found restorers; and in Italy, Platonism was especially illustrated by Ficinus and by Picus Miran-
dolus. The former of these had a powerful influence upon
the Italian Schools of Medicine; and this influence was entirely in favor of pneumatism.* Ficinus held that all matter
in itself is totally inactive, and that in order that any force
may be communicated to it, there must be a nature placed
above it. In the human animal, this seems to correspond to
Plato's mortal soul: it is incorporeal according to Ficinus,
and completely interpenetrates the matter of the body. It
is upon this middle nature, and not upon the material body,
that all impressions and images of external objects are
received.

It would be difficult, perhaps, to estimate correctly the
effect which revived Platonism had upon the progress of
medical theories; but there can be little doubt that the abuse
and corruption of this philosophy had much to do with the
mysticism which tainted medical science for a long period,
and of which Cardan, Cornelius Agrippa, Paracelsus and Van
Helmont were prominent examples, and Rosicrucianism an
incidental result. The great improvements in Anatomy
which were effected by the labors of Sylvius, Fallopius,
Fabricius, Harvey, and many others, were not given to the
world till the latter half of the sixteenth century: indeed,
Harvey's great discovery was not published till 1628. It is,

---

* Vide Buhle. Vol. ii, Pneumatologie de Ficin.

therefore, at an epoch peculiarly favorable to futile speculation upon the deep mysteries of vital action, that Paracelsus comes upon the stage—a true apostle of quackery, though doubtless a sincere enthusiast. He was an illiterate **man ; at least, he** had no regular education, and though he traveled far and wide, and visited many universities, seems never **to have** acquired any **accurate learning, and probably obtained his** notions of philosophy by hearsay only. A strange **jumble** these notions were of Platonic, Aristotelian, and mystical doctrine. The central point of his physiological teaching, or, at least, that point which it most concerns us to notice, is the notion which he propounded of the *Archæus.* This name he gave to an imaginary demon whom he supposed to reside in the stomach, and to govern the processes of nutrition, separating the useful from the poisonous part of the food, and distributing each to its proper destination. This theory will be considered presently, in connection with the doctrines of Van Helmont. Of Paracelsus himself we have no more to · say, than to allude to the important impulse which he gave to chemistry, and also to therapeutics, by his rash, blundering, yet eminently suggestive, experiments, and to the great influence which he undoubtedly exercised on the course **of** medical thought for many years.

Van Helmont was a follower of Paracelsus, and he adopted the notion of the Archæus, which the latter had propounded, and made it the foundation of a host of fanciful speculations. He represented this Archæus as an immaterial force, presiding over all bodily functions: and he subdivided it, giving to each member of the body its own peculiar vital spirit: he taught, moreover, that the *consensus* of all these vital spirits produced health, and their disagreement disease. The central Archæus he placed, as did Paracelsus, in the stomach.

The doctrine of a vital spirit, controlling all the functions of the body, and presiding over the generation of **all the** **organs,** was not a mere abstract idea with Van **Helmont.** **He was an excellent chemist, and an industrious investigator** in this branch of science: among other discoveries, he first established the existence of *gaseous* bodies, and he appears to identify the Archæus itself with a gas. It was, of course, perfectly natural for one who was, doubtless, imbued with Platonic, as we know he was with mystical, learning, to attribute the highest importance to the discovery of a kind of impalpable matter which had so much the appearance of an intermediate nature between body and mind.

Spiritualism runs fairly rampant in the works of this author, and it is sometimes extremely difficult to extract his real meaning from the mass of crude metaphors in which he envelops it. But it is certain that he recognized vital spirits in all* forms of matter, and that he thus established, as did his philosophic teachers, a complete sympathy between the bodily organism of man, and every substance in the universe. He speaks of a celestial and *enormontical* virtue (the very phrase used by Hippocrates) which resides in every organism, and which depends for the regularity of its actions upon an external motor; such for instance, as the orderly movement of the celestial bodies.† And he distinctly recognizes the excitement of the vital spirits by different kinds of medicine, as by "the subductive virtue of cathartic or purgative, the somniferous faculty of hypnotic or dormitive medicaments, &c."‡ These influences he speaks of as "less noble and more corporeal (than certain others), yet abundantly satisfactory to those ends which, by the primitive destiny of their creation, they regard." These words foreshadow the doctrine of *appropriate stimuli*.

Van Helmont was an able and original thinker, and a perfectly honest man; but the tone of his mental organization disposed him naturally to fanatical superstitions of all kinds. The services he rendered to chemistry were very great; but they are almost overweighed by the mischief which he did to

---

* Helmont. Ortus Medicus, p. 43-44.

† A Ternary of Paradoxes. Translated by Walter Charleton, 1650, pp. 33-4.

‡ Ibid. p. 79.

rational physiology and pathology, by the cloud of mysticism in which he enshrouded the philosophy of life and disease. The notion of man's sympathy with the universe, of which he is the microcosm, was elaborated into a multitude of subtle fancies about magnetic influences, and what not: which if any one feels interested in, he may study them in a tractate, "Of the Magnetic Cure of Wounds," which is to be found in the volume translated by Charleton.

The most important aspect of Van Helmont's doctrines is that they form a turning-point, at which purely metaphysical notions of physiology may be seen to mix and blend with legitimate physical science. It might have been expected that this phenomenon would present itself, provided that Chemistry obtained a considerable start over Anatomy in the race of scientific development which was originated by the revival of letters; and thus it happened, quite naturally, that the lucky guesses of Paracelsus, and the far more genuine researches of his pupil, gave a new turn to biological inquiries. The notion of *fermentation* forms the physical element (small as it is) which enters into Van Helmont's theory of inflammation. The morbid cause having insinuated itself into a part of the body, the Archæus (who, it must be remembered, was an actual personality, endowed with intelligence, and with the most lively emotions) became *enraged*, and sent into the parts a ferment which it had always at its disposal, which irritated the tissues, and called the blood into them, and thus became the proximate cause of inflammation. In another place, he says that an inflamed part is in the same state as an organ pricked to the quick by a *thorn*: "it is as if a sharp thorn wounded a nervous tissue."*

The value of this dictum of Van Helmont, as an illustration of the history of the doctrine of Stimulus, is very great. The connection between the *pricking of a thorn* and the Greek word στιγμὸς (a pricking or stabbing), from which our modern

* Van Helmont. Principes de Médecine et de Physiologie. Preface Necessaire. Ort. Med. p. 320.

word "stimulus" is certainly derived, is obvious. The idea of a distinct *injury inflicted*, corresponds well with the notion of the Archæus, a being who would certainly resent such an injury, just as a passionate man might resent a blow. Both Broussais* and Vicq d'Azyr† regard the announcement of these opinions as the real origin of modern ideas as to vital irritability; and the latter expressly grounds his teaching on it. (1792.)

The doctrine of vital spirits continued to prevail. In the first part of the seventeenth century we find Descartes and Bacon‡ each expressing it in his own way. Descartes held that the vital spirits were separated from the blood in the brain, and chiefly in the pineal gland (which he considered to be the center of the brain and the seat of the soul). They are material in their nature, but the matter which constitutes them is extremely subtle, and they are excessively mobile. The source of movements is the contraction of particular muscles, and the relaxation of their antagonist muscles, a greater afflux of vital spirits taking place to the former than to the latter. As to the movements of the vital spirits, they proceed from the great diversity of motions which the organs of the soul experience from the external agents which operate upon them, and from the inferior activity of the principles from which the vital spirits derive their origin. All parts of the body are capable of being moved by the objects of sense, and the vital spirits, without consciousness on the part of the animal.§

Lord Bacon says‖ that "the main differences between animate and inanimate are two: the first is, that the spirit of things animate are all continued with themselves, and are all branched, in veins and secret canals, as blood is; and in living

---

* L'irritation et la folie, p. 10.

† Encyclopédie Méthodique, Art. Aiguillon.

‡ See also Sir Thomas Browne, Common and Vulgar Errors, book iv, chap. 7.

§ Des Cartes, De passione Animæ, i, p. 12. Epist., lib. ii, pp. 36, 38, Dioptr., p. 56. See also Buhle. Hist. de la Philosophie Moderne, vol. iii, p. 17, et seq.

‖ Nat. Hist., Cent. vii.

things the spirits have not only branches, but certain cells or seats where the principal spirits do reside, and whereunto the rest do resort: but the spirits of things inanimate are shut in and cut off by the tangible parts, and are pervious one to another, as air is in snow. The second main difference is, that the spirits of animate bodies are all in some degree more or less kindled and inflamed, and have a fine commixture of **flame and** an aërial substance. But inanimate bodies have their spirits no whit inflamed or kindled. And this difference consists, not in the heat or coldness of spirits, for cloves and other spices . . . . . . have exceeding hot spirits (hotter a great deal than oil, wax or tallow, &c.), but not inflamed. When any of these weak and temperate bodies come to be inflamed, they gather a much greater heat than others have uninflamed, besides their light and heat." It is a curious circumstance that Bacon, though deriding Paracelsus and other "darksome authors," for reviving Pythagorean doctrines, mentions, with apparent respect, the wildest stories of the "natural magic" school, so prevalent in his day.

Helmontian doctrines,* more or less modified, seem to have carried the day in England during the seventeenth century, at least among the theorizers.† The names of Willis and of Glisson deserve some special notice. The former (the great illustrator of the nervous system), besides manufacturing a a new kind of "mortal soul," added something to the theory of vital irritation by suggesting that the blood was provided abundantly with saline principles, which might irritate any part into which it was effused. Glisson, an Oxford professor, and an excellent anatomist, spoke of irritability as a force of which perception and appetite are the factors. Perception precedes the movements which are the effect of irritability, **and is** converted into sensation **as** soon as it is perceived by **the mind.** Perception is inherent in the fibers, it renders

---

* Charleton, the translator of Helmont's Paradoxes, was physician to Charles I.

† It is obvious that the great Sydenham comes into quite a different category; there is little or no place for his name in the history of an abstruse doctrine like that of Stimulus.

them irritable, it is the basis of *natural* motion, as distinguished from *sensitive* motion, the result of a sensation. Irritability is divided into natural, animal, and vital,* and the humors participate in it, as well as the solids. Glisson believes in the existence of vital spirits, intermediate between the soul and the organs.

Toward the end of the seventeenth century three distinguished men arose, who were destined to exercise a powerful influence on the progress of the doctrine of vital irritation. Boerhaave at Halle, and Stahl and Hoffmann at Leyden, were the three leading teachers of their day. The pathological doctrines of Boerhaave had, on the whole, a decided tendency toward mechanical explanations of vital phenomena, as is shown in his theory of inflammation, the principal cause of which he places in the obstruction of the small vessels by the corpuscles of the blood; nevertheless, we recognize in his definition of "disease," as signifying "every condition of the human body which injures the natural, vital, or animal actions," distinct traces of the influence of Galen. Stahl was the founder of the theory of "Animism;" he represented the body as entirely under the domination of the immaterial soul, which alone perceived impressions, and which gave rise to bodily motions by acting on the *tonicity*† of the fibers. Disease consists in the various *irregularities* in the government of the vital economy by the affections of the soul; of its proximate causes the most common is plethora, the great frequency of which affection is accounted for by the universal tendency of mankind to take too much food. Hence arise congestions of the blood-vessels in various parts; these congestions are due to the action of the soul, which excites tonic movements of the vessels, with a curative intention. In the case of the acute fevers these efforts of Nature are ineffectual to produce a cure; greater efforts are then put forth, and

---

* Compare Galen's account of the three forces presided over by the πνεῦμα. *Supra,* p. 44.

† Stahl. De Motu Tonico, &c.

shivering is produced, by exaltation of the tonic movements of the vessels; if even this be insufficient to produce an evacuation of the blood from the parts, pus is generated, and this causes a definite resolution of the inflammation by withdrawing the sulphureous materials from the blood, which gives its redness and heat, &c. Passive congestions, on the other hand, arise from feebleness or spasm of the vessels. Hæmorrhages are means by which Nature often obtains a cure by evacuating the thickened humors.

Hoffmann was educated in the ideas of Van Helmont, and the basis from which he starts is the assumption that matter is in itself totally inert. To account for vital movements, he invokes the aid of the *monads* of Leibnitz, which he makes as numerous as were the atoms of Democritus, and represents them as presiding over every molecule of matter. These monads preside over the development of all organisms, and have the power to alter them or to destroy them, so as to reproduce others, or the same in other forms. There are two kinds of movement in the solids of the human body, contraction and dilatation; these are the cause of all motions of the fluids. The blood is charged with saline, sulphureous and æthereal matters; these are derived from a subtle matter, æther, which circulates in space, and penetrates the body through the lungs and the pores of the skin; it first falls upon the nervous system, and is elaborated in the brain, to vital spirits which are spread all over the body; the lymph of the nerve-sheaths being the vehicle in which it is carried. The æthereal matter, finding its way everywhere into the blood, gives it the power of irritating the solids, and communicating to them the movements of contraction and dilatation. The movements of the heart are thus accounted for; the *impulse* of the blood produces dilatation: the irritation of the muscular tissue by the sulphureous and saline constituents of the blood makes the heart contract again.

The next great step was made by Haller, who declared that the muscular tissue was alone endowed with Irritability.

Nervous tissues, and those abundantly supplied with nerves, are endowed with sensibility; other tissues are without either sensibility or irritability, and possess only a "dead force." Winter, and his followers, however, declared that irritability exists in all parts of the organism, arguing from the evidences of its presence in zoophytes and plants; and, moreover, they distinguished the agents which are capable of operating on it, so as to excite, to diminish, to waste or to extinguish it, by the name of *Stimulants*.*

From this time forward the use of the word "stimulus" is frequent, and the idea of stimulant action begins to assume a prominent place in all theories of life and disease. Cullen was a pupil of Hoffmann, and a great opponent of the humoral doctrines which had been in fashion ever since the days of Hippocrates. He carried to Edinburgh, and established in that school, a doctrine which may be said to be a kind of solidism, in which Hoffmann's notions were combined with the Hallerian doctrine of irritability. Hoffmann had held a theory as to the existence of two opposite conditions, spasm and atony, which at first sight has some resemblance to the *strictum* and *laxum* of the Methodists, whose doctrine, however, he would have rejected as based upon a materialistic philosophy. Both he and his pupil, Cullen, recognized the *vis medicatrix naturæ*. According to Cullen, a state of spasm is produced in the commencement of all fevers; these diseases are uniformly preceded by debility, and the spasm is the result of a reaction brought about by the *vis medicatrix*. Inflammation is caused by irritation of the capillary vessels. Cullen speaks frequently of the action of stimulants, and of the irritation which is caused by them spreading by sympathy to parts of the body remote from that to which they may be applied.

---

* Broussais seems to imply (De l'irritation et de la folie, p. 30) that the school of Winter were the first among pathologists to use the word "stimulant" in regard to vital actions. This may be the case, but at any rate the germ of the idea was in Van Helmont's doctrines.

Dr. James Gregory, the author of the *Conspectus Medicinæ*, gave great prominence to the nervous system in the causation of disease, and discussed the subject of the action of stimulants with care and elaboration. He defined stimulants as agents which excite sensation in the sentient parts and motion in the muscular parts; the majority of them producing both these kinds of action. They operate upon the nervous system: primarily, upon that part of it to which they are applied; and, secondarily, in proportion to the degree of their "diffusibility," upon the nervous system at large. The local result of stimulation is an increased impetus of blood toward the stimulated part, causing redness and heat, often swelling and pain, and sometimes real inflammation; secretion is exceedingly increased or disturbed. Nor are the effects confined to the stimulated parts or to those contiguous to them, for frequently very remote parts which are connected with the former by sympathy, are excited to violent motions, and sometimes the whole body, participating quickly in the affection of a particular locality, is very manifestly excited. In consequence of this, all parts acquire renewed alacrity, vigor, and mobility, and not unfrequently the mind itself exhibits a renewed cheerfulness.

The effect of some stimulants is more, of some less, permanent; but in all cases it is characterized by evanescence, nay, more, stimulation often terminates in its opposite, and the alacrity, mobility, and vigor of body, and the cheerfulness of mind, originally produced, quickly give place to positive depression. The stomach, being liberally supplied with nerves, is an excellent medium by which the influence of stimulants may be diffused over the whole body. **The operation of** stimulants upon the organism is strongest when the latter is feeble; in acute diseases, for instance, **in which the** body **is weak and** unusually **irritated, these diseases, therefore,** require a treatment the reverse of stimulant.*

A more important theory of stimulation than those of Cul-

* Conspect. Medicinæ, cap. 1166, et seq.

len and Gregory was put forward by the celebrated John
Brown, the author of the so-called "Brunonian" system;
which has a peculiar interest for us, inasmuch as it is supposed by those who have not studied Brown's views, that
they were revived in the practice of the late Dr. Todd and
others.   A greater mistake than this could hardly be imagined.   Brown was a theorist *pur sang,* and his system was
as cut and dried, and as remote from the rational empiricism
which has distinguished the present medical generation, and
that which immediately preceded it, as it could possibly be.
Nevertheless it possessed some merits which were by no means
insignificant.   The fundamental idea of the Brunonian system
was,* that animal life cannot be maintained for an instant
without the perpetual operation of stimuli, and that all foods,
and whatever tends to preserve life, are to be looked on as
stimulant in their action.   Every animal is born with a certain definite amount of a faculty called Excitability; the
operation of stimuli upon this faculty produces excitement,
or stimulation, and proportionably exhausts the excitability.
There is a certain *juste milieu* which represents perfect health,
namely when the excitability and the excitement exist in
equal amounts.   If the former be in excess, diseases of debility (asthenic) are produced, if the latter, diseases of over-
strength (sthenic) make their appearance.   Debility is of two
kinds: "direct," when it is the result of a deficiency of stimulation; "indirect," when it is produced by excessive stimulation unduly exhausting the excitability, as in the termination
of sthenic diseases.   Both direct and indirect debility are to
be treated with stimulant remedies.   It is only sthenic diseases, *in their progress toward debility*, that admit of a
lowering treatment, and the number of such diseased conditions is very small, in Brown's classification, compared to
those of an opposite character.

Brown professed the greatest contempt for all systems of
medicine which had preceded his own, yet it is obvious that

* Works of Dr. John Brown; 3 vols. London, 1804.

he was less original than he assumed to be. His faculty of "excitability" is the Hallerian "irritability," but elevated into so distinct a form that, as is justly remarked by Broussais, it takes the rank of an ontological creation; in fact, it is as purely the work of fancy as was the Archæus of Van Helmont. The characteristics of a personality may be traced in it distinctly. But there is one great merit which belongs to Brown, and deserves to be remembered, namely, his appreciation of the fact that disease, in by far the greater number of cases, implies a debilitated condition of the organism.

Darwin, the author of the "Zoonomia," propounded a doctrine which, in its main features, closely corresponded with that of Brown. The following are the "laws of animal causation," as set forth by Darwin:* "1. The fibers which constitute the muscles and the organs of sense possess a power of contraction. The circumstances attending the exertion of this power of contraction constitute the laws of animal motion, as the circumstances attending the exertion of the power of attraction constitute the laws of motion of inanimate matter. 2. The spirit of animation is the immediate cause of the contraction of animal fibers; it resides in the brain and nerves, and is liable to general or partial diminution, or accumulation. 3. The stimulus of bodies external to the moving organ is the remote cause of the original contraction of animal fibers. 4. A certain quantity of stimulus produces irritation, which is an exertion of the spirit of animation, exciting the fibers into contraction. 5. A certain quantity of contraction of animal fibers, if it be perceived at all, produces pleasure; a greater or less quantity of contraction, if it be perceived at all, produces pain; these constitute sensation. 6. A certain quantity of sensation produces desire or aversion: these constitute volition. 7. All animal motions which have occurred at the same time, or in immediate succession, become so connected, that when one of them is reproduced, the other has a tendency to accompany or succeed it. When fibrous con-

* Zoonomia, p. 30-1

5

tractions succeed or accompany other fibrous contractions, the connection is termed association; when fibrous contractions succeed sensorial motions, the connection is termed causation; when fibrous and sensorial motions reciprocally introduce each other, it is termed catenation of animal motions. All these connections are said to be produced by habit; that is, by frequent repetition." Darwin held that the continued application of a stimulus exhausts the "sensorial power or spirit of animation," which obviously corresponds with Brown's "excitability." Perhaps, on the whole, the former term is preferable to the latter, as being more explicit, yet not more ontological, than that of Brown.

But the most decided proof that old influences were still at work, and that pathology had changed in little except in its terms, is to be found in the celebrated article, "Aiguillon," written by Vicq d'Azyr, in the "Encyclopédie Méthodique" (1787). The author expressly adopted the idea of the *spina Helmontii*, already alluded to, as the basis of his explanation of vital stimulation. It is true that he makes no reference to an Archæus, and, indeed, expressly limits himself to the use of such expressions as, "tonic movement," "nervous deturgescence," "muscular movement," &c., and that he refers all the phenomena of stimulation to the nervous system; nevertheless, it is impossible not to see that the metaphysical bias of Van Helmont has extended itself to his commentator. "I conclude," says Vicq d'Azyr, "that in the case where an inflammatory tumor has been produced by a thorn driven into (the spot which now forms) its center, the nerves *excited* by the presence of this stimulating cause, have reacted on the muscular fibers of the arteries; that an increased movement of these fibers is thus occasioned, and a more rapid circulation in the vessels; that consequently an increased quantity of the juices is carried to the part; that the juices are extravasated into the cellular tissue through the dilated extremities of the small arteries; . . . that the blood, thus extravasated, cannot return in sufficient quantity by the

veins, and that these phenomena are not at all more surprising than the fact, that, on our pinching one of its nerves,
a muscle will be thrown into convulsion, or that the same
muscle will distend itself with blood, and contract, when the
internal nervous action directed by volition brings to it a sort
of *stimulus*, the momentary effect of which corresponds to
what I have already described." The reader will doubtless
observe, that this illustration, so far from affording any true
explanation or confirmation of the theory of Vicq d'Azyr, is
itself the expression of a mere hypothesis which has not been
proved. That it should have been so confidently appealed to
by this eminent author is a proof of the reluctance which was
still felt to relinquish the metaphysical method of inquiry,
and the ideas which, by its influence, had become consecrated.

The doctrines of Brown, though counting many able supporters in this country, were by no means adopted with so
general an enthusiasm here as in many parts of the Continent.
It would be quite impossible to sketch, ever so briefly, the
various foreign modifications of Brunonian doctrines; it **must**
suffice to refer to the most notable, namely, to that which took
its rise in the schools of Italy, and found its most characteristic
expression in the teaching of Rasori. This physician was the
inventor of the doctrine of "contra-stimulus," which is so important, that we must examine its relation to the principles
of Brown. The author of Brunonism had stated that there
was, in truth, no such thing as direct *sedative action;* and
that all **the** so-called sedative affections were produced by the
use of a weaker stimulus than usual. Rasori, on the other
hand, maintained the existence of a large class of physiological agents, which possessed the power of producing an
effect on the animal **fiber** directly contrary to the effect of
stimulants, and held that the operation of these agents would
remove the symptoms caused by stimulation, and, if carried
far enough, produce disorders which could only be remedied
by the use of stimulants. He gave to this new class of
remedies the title of "Contra-Stimulants;" and he assumed

the existence of two opposed diatheses in disease—that of
Stimulus, and that of Contra-stimulus, which required respectively to be treated by remedies of a lowering, and of a
stimulant, character. Otherwise, the phenomena of stimulation were explained in much the same way by Rasori, as
they had been by Brown ; but the doctrine of contra-stimulus
led the Italian physicians to a practice very different from
that of Brown ; for they regarded the large majority of diseases as resulting from the diathesis of stimulus, and declared
that, as such, they must be treated on a depressing plan.
Every student of therapeutics is familiar with the history of
the introduction, by Rasori and Tommasini, of large depressing doses of antimony in the treatment of inflammations ; a
practice which may be considered typical of the contra-stimu-
lant method.

Against the doctrines of Brown, there arose, at the commencement of the present century, an opponent of no ordinary
ability and energy—the celebrated Broussais ; whose influence
in his generation was unbounded, and whose so-called "physiological doctrine" rapidly acquired a great sway, the traces
of which are visible enough even now. This author professed
to be the champion of a most important principle. He an-
nounced himself as the founder of a pathology which, for the
first time in the history of pathologies, was based upon an
enlarged and comprehensive acquaintance with anatomy.
The splendid researches of Bichat, which had shed new light
on physiology, by giving an intelligible account of the several
tissues and organs of the body, and of their varied functions,
had excited his enthusiasm, and there is no doubt that he
might have become the founder of a true science of pathology.
Such was not to be the case, however. Broussais did indeed
render important services in opposing the exclusive nervous
doctrines of Cullen and Brown, and in calling attention to
local morbid affections. But he was an inveterate theorist,
and so prejudiced in favor of an idea, that he could not observe
correctly. He took the anatomical system of Bichat, and by

the aid of a lively fancy, built upon it a system of diseases
such as they *ought* to be, and as he therefore thought they
were. The advance of pathological anatomy, particularly in
the hands of the great Laennec, effectually disposed of many
of his statements as to the morbid changes in disease. The
basis of Broussais' doctrine was the assumption that all the
animal tissues are endowed with a property called Irritability,
a property which is called into play by the action of stimuli
of various kinds, and by the operation of which all vital
phenomena are produced. So far he is in accord with
Brown; but the next step in his theory places him in direct
opposition to the latter. For whereas the Brunonian system
places all local manifestations of irritation in quite a subordi-
nate position to the accompanying constitutional disturbance,
with Broussais the first importance is always accorded to the
local affection. To such an extent did the latter carry his
views on this point, that he denied altogether the existence
of "essential" fevers, and declared that all the diseases so
called depended primarily upon an inflammation, simple **or**
complicated, of the mucous membranes of the stomach or intestines.* So, again, pulmonary consumption, which is **now**
well known to be a general affection of the constitution, attended with the deposition of tubercles **in** the lungs, was **to**
be ascribed, according to Broussais, entirely to an inflammation of the mucous membrane of the windpipe and bronchi,
spreading to the tissue of the lung. The doctrine of *sympathy* enabled him to account for the occurrence of inflammation
in a part far distant to that originally affected, in consequence
of irritation conveyed by the medium of the nervous system.
Thus when in the course of rheumatism or gout, delirium or
convulsions set in, it was at once assumed that the inflammation had spread, or had migrated, to the brain. Almost
every morbid phenomenon, with Broussais, was an indication
of inflammation, and inflammation was only a high degree of
the irritation or stimulation which, in its proper degree, was

* **Examen des** Doct. Medicales, &c., vol. i, p. xxxiv

a necessary condition of life. But as it was impossible for a physician of his experience to overlook the fact, that many severe inflammations are accompanied with great general debility, he declared that the development of vital force, *i.e.* of irritation, might be unequal in the various parts of the organism; so that while the vitality of one viscus was greatly exalted, that of the body generally might be much below its normal standard.

The cardinal point of Broussais' doctrine is, as I have already said, that irritation, by which he understands an exaltation of vital force, is at the bottom of every morbid condition; in addition to this, he believed that such irritation always results in the production of an increased flow of blood to the part. The progress of anatomical research has demonstrated the incorrectness of the observations on which the latter proposition was grounded, but the belief that irritation is the necessary starting-point of inflammation, and is itself a proof of exaltation of the vital force residing in the irritated part, or in the neighboring tissues, has survived to the present time, and is expressed in the "Cellular Pathology" of Professor Virchow. We shall return to this subject presently.

It is impossible to think of John Hunter without reverence for his genius and his labors, but it is surely permissible to lament his peculiar bias of mind, or rather, perhaps, his want of early education, when we observe how he has dealt with the subject of stimulation. His language, when discussing this topic, is unsatisfactory.* It is evident that he believed in the existence of something which closely corresponds to the "mortal soul" of the ancients; and it would further appear that he inclined to divide this principle into two, the one presiding over the organic, the other over the sensitive functions; for he repeatedly speaks of the "living principle," and the "sensitive principle," and he ascribes to each of them a sort of intelligence, and power of choice. Nay, more, he even

* Hunter's works (Palmer's edit.), vol. i, p. 268, &c.

goes so far as to endow the body, quite independently of the action of the mind, with a sort of *memory:* for instance, he relates the case of a gentleman who had lost the power of being constitutionally affected by mercury, by long-continued use of that medicine, but who regained his former suscepti- bility to its influence after disusing it for some time; a circum- stance which Hunter accounts for by supposing that the tissues had *lost the memory of the action of the drug.*

Hunter's view of the action of stimuli and irritants makes these agents really appeal, not to the tissues themselves, but to the half-incorporeal principle or principles which reside in them. The difference between a stimulus and irritant is this: the former appeals to and promotes the natural tendency or function of an organ; the latter merely excites the vital or the sensitive principle to action, without increasing any nor- mal function of the irritated part. The excessive action of a stimulus, however, always becomes irritant.

It is probable that the introduction of the vital principles was a measure which rather indicated Hunter's impatience of leaving things unexplained, than any conviction which we ought to receive with that very high degree of deference which is usually to be paid to anything, however difficult to understand, which that great man has recorded. There is one part of his teaching, as to the action of stimulants, how- ever, which throws valuable light on their true nature; that, namely, in which he separates them from irritants, by only attributing to them excitant action of such a kind as tends to promote some *natural* function or tendency. We shall have occasion, at a later period, to reflect on this idea, and we shall find that it contains, probably, the germ of the true defi- nition of stimuli.

Nothing is more remarkable than the immense superiority of Hunter's labors to those of either Brown or Broussais. Each of them was greatly his superior in the logical faculty, and in everything which tends to utilize mental powers which are not of the very first class. But the spirit of his work was true, his power was first-rate, and he had pre-eminently that faculty of insight which Brown and Broussais only possessed in a far inferior degree. Whether or not his individual opinions on some isolated points were correct, is of little consequence to his true glory, which is—that he decided the fate of medicine, and laid the first stone of an edifice which has slowly grown up since, almost unnoticed among the crowd of ambitious but temporary structures which are destined to disappear and leave it standing alone. The science of medicine, thenceforth, was to be established on the study of life as it is in nature, not life as it should be according to the books; and the ultimate result of these studies may be more clearly surmised at present than they could in Hunter's days. One thing which will probably happen is the final interment of his "vital" and "sentient" principles in the tomb of their ancestors.

The epoch which has succeeded Hunter, Cuvier, and Bichat, has been marked by a tendency to cultivate sedulously the minute anatomy of the tissues of the body. The old disputes between solidism and humoralism have not been forgotten, nor have they died out. But the natural consequence of the existence of improved means for the careful examination of the bodily structure has been a less exclusive attention to blood-vessels on the one hand, or to nerves on the other. The observations of Schleiden and Schwann as to the development of living organisms from cells, has given an immense impulse to this movement, and while the more special results of the change are to be traced in the existence of a great school of cellular pathologists, its influence has overflowed on all sides, and has pervaded, more or less, every school of physiology. To Professor Virchow, as the most prominent representative of the cellular pathologists, we naturally turn with interest for an exposition of the doctrine of stimulus as modified by the modern researches in which he has borne so distinguished a part. Contrary to what we should have expected, the views expressed by him correspond more nearly, perhaps, than those of any other physiologist of the day with the older doctrines.

It need hardly be premised that the Cellular Pathology regards the minute elements of the tissues themselves, as the *locus in quo* of vital actions and of diseases.* "Every vital act," says Virchow, "presupposes an excitation, or, if you like, an irritation. The irritability of a part, therefore, appears to us the criterion by which we judge whether it is alive or dead. . . . The different actions which can be provoked by the influence of any external agency are essentially of three kinds. . . . When, namely, a given action is called into play, we have to deal with a manifestation either of the *function*, the *nutrition*, or the *formation* of a part." As an example of the mode of action of stimuli, we may cite the case of inflammation, which according to this author consists in an excitation of the nutritive and formative power of the inflamed part. In a note to the English edition of his work, furnished by himself, he illustrates this process in the following way: "Suppose three people were sitting quietly on a bench, and suddenly a stone came, and injured one of them, the others would be excited, not only by the appearance of the stone, but also by the injury done to their companion, to whose help they would feel bound to hasten. Here the stone would be the *irritant*, the injury the *irritament*,† the help an expression of the *irritation* called forth in the by-standers." Upon this theory of stimulation it may be observed, that it belongs essentially to the same category with the vitalism which in earlier times was constantly invoked to solve all difficulties; and that it appears to be of very doubtful advantage at the present day, when we are constantly learning from the investigation of physical laws to explain some action of our bodies which had previously been unintelligible. That it descends from Broussais, the professed enemy of metaphysical explana-

---

* Cellular Pathology (translated by Dr. Chance), p. 287 et seq.

† The German word used for the passive change in the tissues immediately injured.

tions of corporeal processes, is not surprising, for vital irritation, such as Broussais conceived it, was as much an artificial
abstraction as were any of the vital principles, &c., which it
was intended finally to displace.

Professor Bennett, of Edinburgh, another representative
man among recent pathologists, in the account which he gives
of the early stage of the inflammatory process, speaks of it as
originally caused by an irritation of the ultimate molecules of
the part, in consequence of which their vital power of selection
is destroyed, and that of their attraction is increased. The
strong pulse, the fever, and the increased flow of blood in the
neighborhood of inflamed parts, are the results and not the
cause of inflammation, and show that the economy is actively
at work repairing the injury. The original irritant is either
an overstimulation of the part, or an irritation conveyed from
a distance by the medium of the nervous system. A stimulant which, in a certain degree and manner of application,
would assist the natural nutrition of a part, will, when applied in excessive quantity, produce a destructive effect, in
one sense, although the mere bulk of the part may be increased, and certain phenomena, by which the *activity* of an
inflammation is commonly gauged, may make their appearance. The justice of the view which associates with inflammation a certain tumultuous and indiscriminately increased
formative power of the tissues of the part, is borne out by the
general result of modern microscopic research; of this fact
examples may be cited from the investigations of Redfern[*]
upon abnormal nutrition of the human articular cartilages,
and of Strube[†] and His[‡] upon the normal and pathological
anatomy of the cornea, and from the observations of many
other pathologists. But it is also most clearly established
that the increased formative action is not a mere exaggeration of that which is proper to the healthy state, since its re-

[*] Monthly Journal of Medical Science, 1849–50.
[†] Die normale Bau der Cornea, &c. Würzburg, 1851.
[‡] Beiträge zur norm. u. pathologisch. Histol. der Cornea. Basel, 1856.

sults are of a *low kind*, that is to say, the formed matter is always imperfect. Professor Bennett's practical inference from the natural history of inflammation is, that the results of that affection can only be remedied by the natural action of ordinary cell-processes ; and so far he appears to be justified by undoubted facts. When, however, he speaks of the evils of "overstimulation"* in the treatment of inflammation, he seems to be relapsing, to a certain extent, into the vitalistic notion that it is a stimulant action which produces the inflammatory process.

The practice which the late Dr. Todd made so conspicuous was based on opinions which closely resemble those of Professor Bennett ; with this important difference, however, that he rejected altogether the vitalistic notion of overstimulation. Those substances which are pre-eminently known as stimulants (viz. the alcohols), he prescribed fearlessly in inflammatory diseases, because he regarded them, in proper doses, as foods, and believed that the evils arising from an accidental excess in their medicinal use did not lie at all in the direction of increased inflammation. It cannot be denied that Dr. Todd's practice, so far as it was based on an acceptance of the theory of Liebig (hereafter to be noticed), rested on an unsatisfactory foundation. Meanwhile, that theory served him, as an unsound hypothesis has often served a genuine inquirer for truth, as the means of making most important inductions. The result of his practice was the conviction, in the minds of those who had most closely watched it by the bedside, as carried out by himself, that the idea that internal inflammations could be directly aggravated by the constitutional action of large doses of alcohol was completely erroneous. The influence of his teaching was directly hostile to vitalism.

The observations of Professor Lister must be considered as placing many of the phenomena of morbid irritation in an entirely new light. In his valuable paper on the early stages

* *Vide* Lancet, August 16, 1862.

of Inflammation,* this author demonstrates the following facts: 1. That the early stages of a local inflammation are characterized by a singular tendency of the corpuscles of the blood to adhere to each other, and in this way to produce a continually-increasing obstruction to the capillary circulation. 2. That this adhesive quality of the corpuscles does not appear to be a vital property, since it is never developed in the body in a state of health, but can be artificially produced in blood which has been withdrawn from the body, by means which modify the chemical constitution of the fluid plasma of the blood. 3. That it always affects the white corpuscles more strongly than the red, the vital organization of which is higher than that of the former. 4. That the influence which causes this abnormal adhesiveness of the corpuscles is due to some change in the condition of the blood-vessels which contain them; and that this change is of such a nature as lowers the vitality of the tissues of these vessels. 5. That even the early stages of a local irritation are accompanied with depression of the vitality of the part: and that 6. The general conclusion seems warranted, that inflammation consists in the removal of the modifying influences of the living state, allowing the physical properties of the parts, hitherto restrained, to come into play.

The seat of the changes thus described by Professor Lister is the web of capillary vessels. He shows that the dilated condition of the arteries leading to the part (the cause of the so-called "determination of blood") is a phenomenon apart, and merely incidental to the inflammatory process, instead of being its cause, as has often been supposed; that it depends, in fact, on nervous irritation transmitted from the part to the nervous centers, and reflected thence upon the arteries. These observations certainly afford the strongest argument against the notion that morbid irritation of a part, at any rate from local causes, implies an exaggeration of its normal vital action. All the facts observed seem, on the contrary,

* Philosophical Transactions, 1858.

to indicate that it is the depression of vital powers which produces those diseased actions, the most striking expression of which is seen in a severe acute inflammation. One thing is especially to be noted—that all really irritant action (*i.e.* which is capable, if prolonged, of causing inflammation) is of a radically distinct, if not opposite kind from whatever increases the proper functions of a part, whatever, in fact, makes it to be *more alive*. We have no right, for instance, to speak of the agglomeration of blood corpuscles which takes place when mustard is applied to a delicate web of nearly naked capillary vessels, as a higher degree, or as any degree, of the same action which is produced when we swallow minute quantities of mustard with our food.

The next result of modern research which attracts our attention as bearing strongly on the idea of stimulus, is the discovery by Bernard and Brown-Séquard of the remarkable phenomena attending lesions of the sympathetic nervous system. These physiologists observed that the division of the sympathetic nerve in the neck was followed by an immediate elevation in the temperature of all that side of the face and head which corresponded to the injured nerve; the skin was red and hot, the arteries beat with increased force and fullness; there was augmented sensibility to pain; the eye and ear seemed oversensitive to light and sound, respectively; and the muscles were more easily excited to contraction than usual; the pupil, also, was strongly contracted. These phenomena, and certain others of less importance which accompany them, immediately disappear on the application of galvanism to that end of the cut nerve which is nearest the head. This remarkable discovery has naturally been followed up. In the first place it has been proved by M. Bernard that the arteries of the limbs are controlled in the same way as are those of the head and neck, and that their liberation from this control by section of their vaso-motor nerves is followed by similar phenomena to those observed in the latter situation. It has also been ascertained that paralysis of the sympathetic nerves which accompany the arteries going to secreting glands has a remarkable influence in increasing secretion, and that stimulation by galvanism of the paralyzed nerves temporarily arrests the increased flow.

A most important innovation upon the old ideas of vital action has been recently advocated by Dr. Radcliffe. In the older systems of physiology, motion of **every** kind in the human body was assumed, as a matter of course, to be the result of a property especially belonging to living beings, and **thus it came to pass** that the amount of mobility, or rather of movement, manifested by an animal organ or organism was taken as the measure of its vital force. Dr. Radcliffe has **pointed out,** more forcibly than any other author, that we are not justified in placing so low a value as is implied in this view on **the** influence of **the** physical forces which pervade the universe in the production of the so-called "vital" motions. The most striking **of** all vital motions—those of the muscular system—may at least be strongly suspected to depend upon the physical properties **of** the tissue in which they are developed; and the true action of vital force would appear to be rather that of restraining muscular contraction than of exciting it. This theory is supported by numerous arguments, **of which** some are derived from the superficial, and others from the more recondite phenomena of muscular contraction. Thus, it is notorious that the most steady **and** persistent contraction which muscle can possibly exhibit is that which sets in after death, which does not occur **till the** last remnants of vitality have disappeared, and which, once having commenced, never relaxes till putrefaction releases the tissue from the ordinary laws of molecular attraction. It is well known that sudden hæmorrhages, by which the brain is at once deprived of its accustomed supply of blood, are frequently accompanied or followed by severe involuntary convulsions of the muscles, over which the brain presides by the medium of the nerves. Further, although it **is true that** interruption of communication between brain and muscle, or

spinal cord and muscle, destroys the power of voluntary contraction, yet it is certain that the paralyzed muscle retains
the power of contracting; for not only do involuntary contractions frequently take place, but, after a time, a permanent
state of contraction nearly always sets in, which is closely
analogous to the *rigor mortis*. These and very many other
striking facts appear, even to common observation, to connect
the phenomena of contraction more with the operation of
physical than with that of vital laws. What then is this
vital force which opposes contraction? Are we about to
escape from the old vitalism only to plunge into a new abyss
of metaphysical theory? There is no need for such a proceeding, for there are physical forces which, although we
dare not say that they are identical with the vital energy, at
least are closely related to it, and, indeed, may be individu-
ally shown to be present and active in the animal body. And
with regard to one of these forces, electricity, the progress
of research of late years has given us the most striking proofs
of its importance in the vital processes. The presence of this
agent in the body was first demonstrated so long ago as the
days of Galvani, but it has been reserved for later observers,
and especially for Du Bois-Reymond, Matteucci, and Chauveau,
to develop the history of its action.

In every fragment of the animal body there exist electrical
currents, and the laws of action of all these currents are identical, whatever their situation; the most notable development
of them, however, takes place in *muscle and in nerve*, in both
of which tissues the current may be traced as proceeding from
the ends toward the sides, that is, from the transverse section
to the sides. Numerous experiments tend to show that the
muscular and nerve currents are always weakened during
contractions which occur in life, and that they are altogether
extinguished before the *rigor mortis* sets in. It appears
highly probable that both living muscle and living nerve are
composed of an infinite number of electro-motive molecules;
that whatever disturbs the relations of the electro-motive

molecules of a nerve will weaken its current; that the weakening of the nerve-current will give rise to extra currents of high tension in the nerve, and to induced currents in the neighborhood of the nerve, and consequently in the muscle to which it is distributed; and that these induced currents will so disturb the electro-motive molecules of the muscle, as temporarily to suspend *its* current, and thus allow the natural attraction of its molecules to come into play and cause contraction. As regards the action of these substances which especially possess the power of provoking muscular contraction by their influence upon the nervous system, it is interesting to refer to the case of strychnia, the most typical of these drugs. The researches of Dr. Harley have rendered it extremely probable that strychnia acts by preventing the oxygenation of the blood, in which case it can hardly be supposed to communicate increased force to the nervous system. Rather it would appear probable that it reduces the muscles to a condition in which they obey the laws of inorganic matter.

Dr. Radcliffe inclines to place **the** phenomena of pain on a level with those of involuntary muscular contraction, as regards their vital significance. Both these phenomena will be discussed hereafter, in their relations to the action of **stimulants;** at present we need only remark that, although the new opinions as to their causation are not as yet so established as to command universal assent, they are supported by a mass of evidence which calls loudly for a revision of the older ideas.

One of the most noteworthy features of the state of medical opinion of late years, is the prevalence—which has been increasing ever since the downfall of Broussais' theories of disease was accomplished by the efforts of the anatomical pathologists—of free empirical inquiry as to the action of remedies **upon the body. The old contest** between dogmatists and empirics, which commenced in the days of Hippocrates and his successors, has been going **on,** more or less, ever since; **but in** these later times the empiric is a man of real knowledge, who does not neglect, as did his prototypes, the sciences of anatomy and physiology. The effect of the reaction against Broussais' doctrines has been a general tendency *to specialize* diseases, and to seek for remedies which may annihilate them at a blow, for antidotes, in short.*   The result of this has, in one sense, been disastrous for therapeutics; for it has thrown that branch of science even more decidedly behind the physiology and pathology of the day than it would otherwise have been. That is to say, although individual discoveries which have been made are highly valuable, the therapeutic art has not been regenerated. As soon as a medicine has been proved to have curative powers in a certain number of cases, its *modus operandi* is somewhat carelessly assigned to some.one of the recognized classes of medicinal action, instead of being made, as it should be, the point of departure for fresh experiments, which might help to test the accuracy of the classification itself, and of the physiological theories on which it depends. What do we know for instance, or what did we know till quite lately, of the real physiological action of strychnia, one of the most remarkable medicines which modern practice employs? We assumed that it was an irritant to the spinal cord, but this was merely in deference to preconceived ideas as to irritability, so far are modern empirics from throwing off the yoke of theory; and, in fact, the researches of Dr. Harley were the first scientific attempt to give an explanation of the matter. The main feature of interest to the empiric discoverer of a remedy, is, that it will cure such and such an affection, and he is apt to be careless about its physiological relations to the organism, or to accept a hasty explanation of them. And in this way it may happen that an erroneous physiological theory derives a fictitious support from its supposed applicability to the case of a particular medicinal agent.

That it is now high time for this incomplete method of inquiry to be exchanged for a better, there can, I think, be

---

* **Trousseau** and Pidoux. Traité de Thérapeutique. Introduction.

little doubt. For what is the result of our investigation of
the history of the doctrine of stimulus? Surely this—that
it originated in hypotheses, most excellent and useful in their
day and generation, which ancient philosophers adopted for
the purpose of helping themselves some little way further
toward the comprehension of things which it was impossible,
in the nature of things, that they could fully understand. So
prodigious was the force of mind in these great men, and so
vividly could they portray that which they wished their immediate audience to see and understand, that the very wisdom
and beauty of their words have proved, in one sense, a serious
misfortune to mankind. For the large number of the com-
mentators who followed the golden age of philosophy, seem
to have been smitten with a double blindness; first, in not
knowing what their great teachers actually said, and secondly,
in being unable to perceive that they did not, in many cases,
mean it literally at all, but only figuratively.* From this
double process, first of mutilation and next of incrustation, it
has come to pass that old ideas, professedly tentative, have
assumed their place as fundamental truths in medical belief,
and have lasted to this day. It seems to be unquestionably
to this source that we owe our ideas of stimulation, excitability, irritability, &c., for all these words refer to the residence in the tissues of the body in some essence which,
although not exactly mind, possessed at least passions and
emotions and volition, and to which stimulants and irritants
appealed. If this be the case, there can be little doubt that
it is now desirable, in presence of the facts of modern physiology, to make a *tabula rasa*, and to sweep such ideas
altogether from the field. If, as seems indeed unavoidable,
the words *stimulate, excite, irritate,* and their derivatives,
are to be retained in use, we must at least seek for some new
definitions and limitations of their meaning, for it is unphilosophical to speak as if we still believed that a demon or demons

---

* The combination, or confusion, of Platonic, Aristotelian, and Galenic ideas in
Philoponus is a notable instance.

resided in our stomach, our nerves, or our ultimate cells, whom we could propitiate with cordials, soothe with anodynes, excite with stimulants, or inflame into active wrath with irritants.

It may appear to some, that to ignore the participation of spirit in every action of the living body, would be to give a materialistic turn to our ideas of life. Surely, however, the very contrary of this would take place. If we could shake ourselves free of the old hypotheses of vital action, we should be far better prepared to realize the true character of mind— changeless, eternal, unconfined to time or space—and we should hardly read such language as we occasionally see in medical works, language which would almost imply that intellect is a secretion from the brain. Many of us must have felt the great impropriety of such expressions as are constantly made use of in describing the action of the so-called inebriants; it is quite customary to speak of these agents as "stimulating the mental powers" in their first operation. Again, it is a common thing to read, in works on insanity, of measures calculated to "reduce the mental excitement." The influence of the constant use of such expressions must doubtless be injurious to the physician's judgment. The assumption that particular mental states can be induced or removed by the production of what we choose to consider analogous corporeal processes, is unfounded and mischievous. All that we can know is, that the material organism of the brain should be placed in that state of nutrition, &c., which represents material health, and that when this has been done, the best chance has been afforded for the mind to act rightly. Of the influence on the mind which particular variations from this state will produce, we know nothing accurately. If any one will analyze for himself the train of reasoning by which the treatment of acute mania by large doses of opium must have been arrived at, he will find it highly unsatisfactory. "Acute mania is a state of mental excitement and derangement, characterized by want of sleep. Opium is a medicine which is calculated to produce sleep, and to relieve pain, which is an excitement and derangement of the nervous force. Therefore opium, in the doses calculated to produce these effects, is the proper remedy for acute mania." Such is the kind of argument by which the routine practice of giving large doses of narcotics in acute mania has been supported, against the manifest testimony of experience, which is now, however, gaining a tardy victory. That improvement has been so long delayed is, doubtless, owing to the inveteracy with which we have clung to the doctrine of stimulus, which has been a sort of central point from which the mind could constantly launch out into new excursions into the domain of theoretical pathology.

I have reserved to the last the consideration of the opinions of recent writers on Therapeutics, as to the action of stimulants, that we may be enabled fairly to judge how far this branch of medical science has kept pace with modern pathological research. I propose now to examine briefly the accounts given by the authors of several therapeutical textbooks, which doubtless represent the general way of thinking, as to the action of stimulant medicines.

Dr. Pereira* classes stimulants under the general head of *Neurotic* remedies (acting upon the nervous system), and under the special class of Ganglionics (acting upon the sympathetic nervous system). He adopts the definition of Dr. Billing—" A stimulant is that which, through the medium of the nervous system, increases the action of the heart and other organs, by calling forth the nervous influence, or by facilitating the extrication of it in them." The local action of stimulants is not necessarily accompanied with any change, either chemical or anatomical, although some of these agents do produce a chemical change in the parts with which they come in contact, but which is not an essential feature in stimulant action. Swallowed in moderate quantities, they convey a sense of warmth to the stomach, promote the contraction

* Materia Medica, vol. i, p. 222, 4th edit.

of the muscular coat of the stomach and the intestines, and thereby expel gaseous matters, and assist digestion; and in general they produce hyperæmia and increased secretion of the mucous follicles of the stomach and intestines. In larger quantities they excite thirst, and give rise to nausea or vomiting. The active principle of most, if not of all of them, becomes absorbed; in many instances with great rapidity; and a large number of them are again rejected from the body, through the secretions, unaltered. In the blood they act as stimulants to the heart and blood-vessels, and in passing out of the system in the secretions they act as stimulants to the glands. They cause increased quickness of respiration, and increase the temperature of the surface; and they stimulate the brain and spinal cord to a more active performance of their functions by causing them to be more copiously supplied with blood. In this way stimulants, by causing an increased supply of blood to the various parts of the body, act physiologically as functional exalters, or pathologically as exciters of a febrile state. Their medicinal uses are as follows: 1. As local *acrid* applications. 2. As stimulants to the gastro-intestinal canal. 3. As stimulants to the heart and vascular system. 4. As stimulants to the brain: to arouse the energies, and correct certain disorders, of the nervous system, particularly certain *spasmodic* affections. 5. As stimulants to the secreting organs, they may increase, or alter, or in some cases check, the secretions.

Dr. G. B. Wood,[*] one of the authors of the "United States Dispensatory," describes stimulants as agents which exalt the forces of the system. "A property common to all stimulation is, that it is followed, in the ordinary state of the system, by a degree of depression bearing some proportion to the previous excitement. There may be conditions of temporary prostration in which stimulants may put the system in the power of resuming its ordinary grade of action without subsequent depression, but these do not come within the

* Treatise on Therapeutics and Pharmacology, p. 48.

general rule. The depression is dependent upon the temporary *diminution of excitability* resulting from excessive
action. If the stimulant influence be continued, it follows,
as a consequence of the diminished excitability, that a greater
amount of the stimulant agent must be employed to produce
the same effect, and the excitability is thus still further diminished, until in the end the system entirely refuses to
respond to the ordinary healthy excitement, and morbid and
often fatal debility results. This is an evil against which it
is necessary to be constantly on our guard in the use of stimulant measures. Another, scarcely less important, is the production of inflammation by the excessive or repeated excitement to which the stimulated organ is exposed." Dr. Wood
divides the general stimulants (which affect the whole system)
thus: the lowest grade of action he assigns to Astringents; a
somewhat higher to Tonics; and the highest to the Diffusible
Stimulants, which powerfully and rapidly affect all parts of
the nervous system, those which minister to Animal, as well
as those which minister to Organic, life.

Dr. Neligan* defines stimulants, in general terms, as
"agents which produce a sudden, but not permanent aug-
mentation in the activity of the vital functions." He dis-
agrees with the common opinion which ascribes their effects
primarily to the nervous system, and believes, from the
effects which are often observed to follow their use, that they
are allied, in some respects, to tonics, differing from the latter
in the greater transiency of their effects; for their stimulant
operation, he says, is "almost invariably followed by a corresponding depression of vital power." Their effects also are
more manifestly perceived by the senses than are those of
tonics.

It will be observed that the three writers above quoted,
though differing in some important respects, agree in ascribing
to the action of stimulants an exaltation of a vital or of a
nervous force which already exists somewhere in the organ-

* Medicines, their Uses and Modes of Administration, 4th edit., p. 318.

ism; and it may be added that they agree also in stating that the first stimulant action is followed by a recoil—a depression. There is a vagueness, however, in the terms in which this *recoil* is described which may well attract our attention. It is nowhere expressly stated by these authors, that *as a rule* the secondary depression is carried so far as to place the vital energies at a level *below* that at which they were when the stimulant was taken; yet unless we know whether *this* was meant, the mere fact that "subsequent depression" always occurs, is of little or no practical value; since the whole question of the restorative properties of stimulants is left in complete uncertainty. Let us take first the supposition that the depressive recoil is *greater* than the previous excitement, and try to work out the problem of the continuous "stimulation" so frequently had recourse to in fevers. A first dose of the stimulant having been given, and having produced the highest degree of exaltation of which it is capable—that is to say, having used up as much vital or nervous force as it can educe—depression sets in: in order to prevent the development of this state we repeat the stimulant medicine, thereby using up fresh vital force. Presently comes the recoil, which we again immediately arrest by a fresh dose, and so on for days, it may be for weeks, together. Now one thing is pretty clear, if stimulation means the calling forth, that is, the getting rid of, a certain quantity of a force already existing in the organism, either the accumulated stores of this force must be immense, or they must be simultaneously repaired by that which can *create* force, or the vital power must after a very short time become completely exhausted, and the patient, whether cured of his fever or not, must be "improved off the face of the earth." The simple facts of the case, however, are these: A fever patient often lies for days taking an entirely insufficient quantity of common food to create new force in place of that so lavishly wasted by the stimulant medicine, possibly taking no such food at all. Yet at the end of a certain time, it may be many days, it is discovered that the necessity for the stimulant medicine has ceased or diminished, not by the patient's death, but by his having recovered his energy so far as to digest common food. Nor is it possible to get over this difficulty by supposing that the tissues have been consumed to such an extent that force has been generated at the expense of bodily bulk; for the patients are comparatively very slightly emaciated on convalescence, if they have been treated in this way. Has there been **great** emaciation, so to speak, of the *nervous system*, **though this has** been invisible to our eyes? It does not appear so, for the patients recover intelligence, sensation, and voluntary movement, with great rapidity. In short, on the **theory** that stimulant action **is** followed by a more than equivalent recoil, we are entirely unable to explain these **facts.** But **if** we agree to say that stimulant action is, **indeed, followed by a** recoil, but that the latter is not greater, **or not so great, as the original** exaltation—the statement loses all importance, since this is exactly what happens *after the digestion of a true food.*

It **is** scarcely possible that so obvious a difficulty has not been perceived by the **many** excellent **authors** who held the opinion which **we are** discussing. The way of escape, which **they probably** found for themselves, is indicated by the fact, **that they believe (*vide* Dr.** Wood *loc. cit.*) that a repeated **excitement of an organ, or of** the system, may cause inflamma-**tion.** This would seem to show, that they assume the existence of **a vital principle, which,** when closely examined, **differs,** in no essential respect, from the Archæus of earlier times. **It** is **capable** of being excited, soothed, or exhausted: it is prompt in action, and violent in its resentments; it is, in short, **a** "spirit of animation."

One **of the** greatest difficulties which the ordinary theories of stimulus **have** to encounter is, the explanation of **the** action of the so-called Antispasmodics. These substances form a **sub-class of** stimulants **in all works of** Therapeutics with **which I** am acquainted : **but, in truth, it** would be difficult to name a single general stimulant which does not possess more or less antispasmodic power.   But if nerve-force be the source of muscular contraction (of which spasm is only a variety), and if the constant action of stimuli is to increase nerve-force, how can we explain the fact of stimuli acting in this way? Dr. Headland says that spasms are due to a fault in the *nervous polarity*, commencing generally in the brain or nerve-centers, but does not explain how stimuli, which are supposed to increase, and even to create, nerve-force, can rectify this faulty polarity.   Dr. Pereira says that the cerebro-spinal stimulants, and especially the Alcohols and Æthers, exalt the power of volition by their action on the brain, and enable the will to control involuntary muscular contractions.   It is evident, however, that such an explanation does not apply to the spasms which are the most frequent of all, those of the muscular coats of the intestines.   How can the will be exalted so as to control a colic?

A striking instance of the confusion of ideas which prevails with regard to the action of stimuli is afforded by the usual descriptions of the operation of chloroform, when inhaled. The muscular spasms which not unfrequently occur shortly before the full effect of the anæsthetic is induced, are constantly spoken of as symptoms of "excitement."   Hereafter it will fully be shown that such an idea is entirely erroneous; and reasons will be given for considering the symptom in question as something very different in its significance.   It is the fact that, so far from being any evidence of excitement, muscular rigidity under chloroform is one of the regular symptoms of advanced chloroform-narcosis with many patients.   It is characteristic of the moment when communication between the nervous centers and the muscles is about to be cut off entirely, owing to the temporary death of the nerves.   If the inhalation be steadily pushed, this lingering influence will be quickly abolished; the muscles will be left to their own vital conditions, and perfect relaxation will result. This is the state which is required for the performance of a

surgical operation. That the previous stage of rigidity does
not occur in many persons at all is probably due, as will be
seen hereafter, to differences of the muscular system in different individuals.

It may be confidently affirmed, that **a stronger instance
than this** of a phraseology which conveys an entirely decep-
tive impression can hardly be imagined. Nor is this a barren
error, the consequences of **which are** trivial; for, as it will be
shown hereafter, this mode of regarding convulsive action
produced by chloroform is productive of the most mischiev-
ous results by the influence it has on the method of administration.

But it is in the neglect of any precise observations on the
effects of *dosage* in modifying stimulant action that the most
serious errors have arisen, as I shall hope to show in my
chapter of general conclusions.

In the ensuing chapter, certain phenomena, which are usu-
ally assumed to be conclusive proofs of a stimulant action on
the organism, will be considered, with a view to test the propriety **of such assumptions.**

CHAPTER II.

WE are now to consider certain phenomena which are ordinarily assumed to furnish decided proof of the antecedent operation of a stimulus upon the organism, and to examine the justice of this assumption. The phenomena in question may be classified under six heads, according as they refer to the mind, to sensation, to muscular motion, to secretion, to circulation, or to nutrition.

I. Every rapidly increased manifestation of mental activity from an external cause is ordinarily assumed to result from the action of a stimulus.

II. All increased sensibility, and all pain, is referred to a similar cause.

III. All convulsive muscular contractions are supposed to result from stimulation, and the more violent the contraction, the more powerful the stimulus is supposed to have been.

IV. All marked increase in secretion is assumed to be due to stimulation of the secreting gland, either direct or through the medium of the nervous system.

V. Every increase, whether of the force, or the frequency, of the action of the heart, or arteries, is supposed to be the result of stimulation.

VI. All increase of the nutritive or of the formative activity of a part of the body is supposed to be due to the action of a stimulus.

I. *Mental Phenomena.*—In order that the following remarks may not be misunderstood, it is necessary to observe, even at the risk of uttering a truism, that the only influence

of physical agents upon the mind, **of which** we are cognizant,
is that which they exercise indirectly, by changing the condition **of the brain**; making the latter more or less efficient
as an instrument of the mind, or of some part **of it.**

In this limited **sense,** whatever relieves the brain from
fatigue, and gives a new capacity for mental exertion, may be
called **a** mental stimulus.   Further, it may be allowed that
whatever increases the physiological perfection **of** the brain to
a limit which, though not ordinarily reached, is still congeni-
tally possible to that organ, is in the same sense to be looked
on as a mental stimulant.   The common idea of the cerebral
stimulant, however, implies more than this.   It implies that
a physical agent may produce, at once, and without the introduction of improved cerebral nutrition, an exaltation of mental activity which **would not** be possible without the use of
extraordinary means.

One of the most obvious considerations which presents itself in an inquiry of this kind is, that we ought not to assume
any extraordinary agency to account for facts which are
capable of explanation in a simpler **way.**   Looking, therefore,
to the great improbability, *à priori*, of a direct instantaneous
increase of the mental powers from the action of a physical
stimulus, we are led to ask whether the increase be not merely
apparent, and owing to the removal of a pre-existing obstacle
to development?   There are strong grounds for supposing
that this is the case with regard to the action of the "cerebral stimulants."   The faculties of the mind are not homogeneous, nor such as are necessarily put in motion all at the
same time, or by the same causes.   On the contrary, nothing
is more certain than that a natural opposition exists between
the reason and the will on the one hand, and the emotions
and the appetites on the other.   We may not hastily assume,
then, that activity of the former implies activity of the latter, or *vice versâ;* we must carefully examine the phenomena
of the so-called cerebral excitement, with a view to discover
what sort of mental activity is developed.   Let us take the

case of Alcohol, Opium, and Hashish, respectively, these three substances being typical of the "inebriant" medicines.

The early phenomena of alcoholic intoxication usually wear an appearance at first sight much resembling excitement. But on analyzing the symptoms we are at no loss to perceive that it is the emotional and appetitive part of the mind which is in action, while the intellect, on the contrary, is directly enfeebled. There is no exception to this rule, that in proportion to the degree in which the lower and more animal nature obtrudes itself in the actions and words of a drunken person, the less of intellectual activity does he display. It is at least possible, then, that the violent outbreak of the passions is due, not to any stimulation of them, but to the removal of the check ordinarily imposed by reason and will. This possibility we shall find converted to something like certainty when we discuss the narcotic effects of alcohol.

With regard to opium there is more difficulty in coming to a decision, because the mental phenomena which are caused by its use are less familiarly known. In the great majority of European constitutions, opium produces nothing resembling mental excitement; the effect on myself, for instance, of a large dose, is mere depression and misery. But with most Orientals, and with some Europeans whose constitutions or whose habits of life are peculiar, a condition is produced by the taking of a large but not fatal dose which is very remarkable, and very difficult to analyze. These persons are able, sometimes without any previous practice, to take large quantities of opium without suffering stupefaction; on the contrary, they appear much exhilarated in spirits, and their minds work with much freedom; in some cases, muscular power and the disposition for exertion seem to be increased, but more frequently there is great indisposition to locomotion or hard work of any kind. These effects last for a period varying from eighteen to forty-eight hours; they are succeeded, in some cases, by a heavy, semi-comatose sleep of long duration; in other instances no particular after-effects are noted.

It is impossible, without far more precise data than are at present accessible, to determine the true meaning of these facts. Here, again, we are bound in the first instance to take the simplest explanation possible, and to see how far this can be applied to the phenomena in question. This may be found in the supposition that, in the absence of any extraordinary circumstances, the apparent exaltation of certain faculties should be ascribed rather to the removal of controlling influences, than to positive stimulation of the faculties themselves or of the physical machinery by which they work. It remains to be proved (and a rigorous proof ought to be required), that in circumstances of ordinary health and nutrition, large doses of opium (proportionably to the taker's idiosyncrasy) can cause *general* exaltation of the mental activity; indeed, the details of every such case with which I am acquainted, when closely scrutinized, have seemed to give evidence of a really poisonous and depressing influence operating upon some portion of the brain simultaneously with the apparent stimulation. Something of the nature of paralysis, though it might be partial, has always appeared to me to characterize the action of large doses of opium from the moment that they have completely entered the circulation, supposing the state of health and nutrition to be normal at the time, and the circumstances of climate, &c., such as we know them in this country. And it would seem reasonable to account for the apparent increase of intellectual force, and of emotion, on the supposition that other activities are suspended, and their interference removed. For that would be a violent and improbable theory which should suggest that one portion of the nervous system was excited, and another paralyzed, at the same time and by the same agent.

The "exciting" influence of hashish (and other preparations of Indian hemp) appears to be exercised on the emotions and passions, and on the fancy. Great exhilaration of spirits, of an unreasoning character, is the chief feature in most cases; in some, the passions rage furiously; in others, the chief phenomenon observed is the involuntary production of fantastic mental images, of which I shall have more to say under the head of Narcotism. After a careful comparison of the accounts given by Von Bibra, Moreau, De Lucca, and other writers, with those which I have received from various persons who have eaten hashish, either in this country or in the East, I am persuaded that the above effects ought not to be called stimulant at all; that they are, in fact, the results of a partial and highly peculiar kind of paralysis of the brain. The *true* stimulative action of Indian hemp is a very different thing, and is produced by doses far too small to "excite" the emotions or the fancy in the way which occurs in a regular hashish debauch. Such at least is my present belief, but I am even now carrying on fresh experiments to test the matter thoroughly.

The above considerations are surely such **as should warn** us not to use the phrase "mental stimulation" **indiscriminately**, since they throw so much doubt on the exciting character of three substances which are usually **cited as** typical "stimuli of the mind." And an additional argument to the same purpose may be found in the fact that substances like ammonia, which are universally allowed to act as pure stimulants, are in no case chargeable with the production of any such phenomena as those which are supposed to indicate mental excitement, although they may restore consciousness where it has been temporarily lost owing to cerebral weakness.

II. *Phenomena of Sensation.*—It is usually assumed in physiological discussion, that all pain, and all increased sensibility is an indication of exalted function of the sensory nerves, or of the sensory tract of the nervous centers, produced by the operation of a stimulus.

In order to examine so complex a phenomenon as pain with any chance of success, it is necessary first of all to direct our attention to its simpler varieties.* For many reasons, Neu-

---

* For a remarkable anticipation of the views on the nature of pain which appear to be deducible from the most modern observations, I may refer the reader to the doctrines of Plato and Aristotle, as set forth in *Macmillan's Magazine* (October, 1863).

ralgia, limiting itself to the track of a single nerve, the course of which can be easily identified, is the most typical kind of pain which can be selected; and chiefly, because it is at once distinct in character, and involves the minimum of organic change in the affected part. There can be little doubt that in such an affection as this, so far from stimulation being to blame, the whole circumstances point to lowered vitality as the antecedent cause.

On this topic I am able, unfortunately, to speak with the authority of personal experience. I have been subject, ever since childhood, to severe attacks of neuralgia of the brow, entirely independent of digestive derangement, although liable, as this affection usually is, to be aggravated by any coincident stomach disorder. The pain always follows accurately the course of those branches of the fifth nerve which are distributed to the forehead, the internal angle of the eye, and the nose, more rarely extending also to the branches derived from the second division of the fifth nerve, and distributed to the cheek, *but always on the right side only.* This kind of headache began to trouble me at about the age of fourteen, and for two or three years was of frequent occurrence; for many years past, however, it has been an infrequent visitor. One circumstance ought to be mentioned, although I cannot pretend to estimate its exact relation to the production of the neuralgic pain, viz., that about the time when the headaches first occurred with any severity, I began to suffer from an obstruction of the lachrymal duct on the same side. This obstruction has been ascertained to depend upon a tight stricture of the upper end of the nasal duct, close to the lachrymal sac, and is apparently caused by a tough fibrous cicatrix, probably the relic of some past ulceration. Treatment by the passing of metallic probes has been adopted from time to time, with great temporary relief, but the obstruction has always recurred, and, as a consequence of it, the discharge of tears from the gland is inconveniently profuse. The attacks of pain are invariably caused by fatigue of body or mind, and are preceded and accompanied by pallor of the face, weak pulse, and a general sense of depression. The only remedies which are of the slightest value are sal-volatile, hot tea, and, occasionally, quinine in a full dose, or a glass of wine. **Ammonia and hot tea are the most frequently useful.**

In such a case as this there is little reason to think that the function of the painful nerves can be in excess. The general bodily **energy is always** at a low point when the attacks occur, and the nerves of the part are habitually **in** circumstances which must tend to lower their functional activity. The constant passive flow of tears from the gland, and the curious fact that the hair of the right eyebrow years ago became decidedly gray at a point exactly opposite the supra-orbital nerve, are so many indications of *defective* nervous energy, and are corroborated by own sensations when free from pain, for I am conscious of defective power in the orbicularis muscle, and of a somewhat blunted common sensibility of the skin around the inner angle of the **eye.** Under these circumstances it is, that fatigue being superadded (especially when accompanied with longer abstinence **from food** than usual), an attack of acute neuralgic pain supervenes. The conclusion appears inevitable, that the pain is here the direct consequence of a further depression of an already feeble vitality in the nerves. There is one other most effectual remedy besides those already spoken of, namely, the application of blistering fluid to the skin immediately over the painful nerves; a remedy which, from its inconvenience, has been but seldom adopted. Now, the effect of this must certainly be to increase the supply of blood to the painful nerve, and, *pro tanto*, to heighten its vital energy.

Dr. Radcliffe, in his lectures* delivered not long since at the Royal College of Physicians, has given a description of his own experience of facial tic-douloureux, which in every way coincides with what I have observed. "About thirteen

---

* Lectures on Certain Disorders of the Brain and Nervous System, &c. See *Lancet*, February to July, 1863.

or fourteen years ago," says Dr. Radcliffe, "I had an opportunity of becoming acquainted, in my own person, with the history of facial neuralgia, or tic-douloureux. At that time I was anything but strong or well, and the pain, so long as it lasted, did not tend to improve my condition, for it took away my appetite, and kept me awake at night. For the first two or three days after the commencement of the attack, the painful part of the cheek would admit of pressure, and the face was pale and perspiring. I was dejected also, and troubled with frequent chills and shivers. On the third or fourth day the painful cheek would become swollen, hot, and tender; a state of general feverish reaction would be developed; and, contemporaneously with these changes, the true neuralgic pain would come to an end. I had abundant opportunity for knowing that this was the true order of the changes: first, neuralgia without local tenderness, swelling, and redness, and with frequent chills and shivers, and a decidedly depressed condition of the circulation; and afterward, local tenderness, redness, and swelling, and general feverish reaction, without chills and without neuralgia. I also find that my own experience in this matter is the exact counterpart of the experience of several patients who have come under my notice at different times."

The neuralgia from which I have personally suffered has not occurred in so severe a form as this, nor have the attacks lasted so long (except on one occasion), but the course and termination of the symptoms closely resemble those which Dr. Radcliffe describes. I have particularly noticed the great tenderness which supervenes in the affected parts as soon as the neuralgic pain subsides. I suspect this depends on congestion of the parts.

The history of neuralgia generally, as observed on the large scale in the out-patient room of a hospital, substantially confirms the idea of the pathology of this complaint which is suggested by the experience recorded above. Neuralgic complaints among the poor and laborious classes are numerous,

and they present three varieties, speaking roughly: 1. Neuralgia of the head and face. 2. Neuralgia intercostalis. 3.
Sciatica. The great majority of the patients who apply for
relief from either of these maladies are anæmic, and in a fair
sprinkling of cases the anæmia is obviously connected with
ague-poisoning. The most speedy way of obtaining a temporary relief from the pain in all these cases is certainly the
application of a local stimulant, more especially of some powerful volatile agent such as mustard, or, still better, chloroform diluted with seven parts of some simple liniment. Still
more effective, though, from its nature, slower in acting, is a
blister. The more profoundly the general health has been
affected, and especially the greater the degree of anæmia, the
more necessary it becomes to join with the use of stimulants
(both local, such as above mentioned, and general, such as
the carbonate or muriate of ammonia, taken in five and ten
grain doses respectively), a treatment directed to improving
the quality of the blood. The great efficacy of quinine in
aguish cases is easily accounted for; the few cases in which
it succeeds which are not traceable in any way to the malari-
ous poison, owe their cure, I strongly suspect, simply to the
remarkable effect which quinine certainly has in improving
the systemic circulation: an action of which I have seen many
curious proofs, though it is by no means exerted equally in
all subjects. For the general run of chronic non-malarious
cases, the most successful plan is the use of food tonics to the
blood and nervous system, such as cod-liver oil,* arsenic or
steel (the first of these being incomparably the best), or a
combination of some of them, joined with the use of local
stimulation by means of frictions with dilute chloroform, and
the manipulations of the scientific *shampooer*. And even
among the better classes, whose diet, &c. is good, a modification of the same treatment is still far more often successful

---

* Dr. Radcliffe reports favorably of the hypophosphite of soda in neuralgic cases,
and believes that the phosphorus of this salt (which is but loosely combined) acts as a
direct food to the nervous tissue.

than any other. The subcutaneous injection of small doses
of morphia is a plan which I have often adopted for the purpose of giving temporary relief to a patient who was obliged
to go through his ordinary day of labor. I only use such a
small dose ($\frac{1}{8}$ to $\frac{1}{4}$ grain) as to produce a general stimulant
effect, without the least drowsiness.

The next variety of pain is one of which we may be said to
owe our knowledge to Dr. Inman. I refer to the complaint
which that author has termed Myalgia. The pain in these
cases is referred to the muscular structures, and is often present in its greatest severity in the tendinous ends of those
structures. Of the existence of a malady with such characteristics there can be no doubt in the mind of a competent
observer, whose attention has once been directed to the subject; but the puzzle involved in the facts is great, seeing that
the tendinous structures of the body are very slenderly, if at
all, endowed with nerves, and even the muscular tissue would
seem to be poorly gifted in this respect. Whatever the explanation may be—whether or not we are to believe that
pain may originate in non-nervous parts, and be conveyed to
the percipient brain by the nerves of the neighboring tissues
or not—certain it is that myalgia is a most frequent complaint, and that it attacks only such muscles *as have been
overworked proportionably to their nutrition.* Here again,
the most speedy remedies are such as improve the circulation
of the part without fatiguing it. But both in myalgia, and
also in all forms of neuralgia, the most permanent relief is
obtained by means which gradually increase the general
nutrition—arsenic, steel, cod-liver oil, and the like; and, pre-
eminently, by a highly nourishing diet.

In the pains which accompany some varieties of inflammation we meet with what at first sight seems a complete contradiction of that order of phenomena in the causation of pain
which we have just been considering; namely, pain occurring
in organs which, from the nature of the case, must contain,
on the whole, more blood than usual. But the phenomena of

painful inflammations may be explained by dividing them into two classes. 1. Those in which pain occurs at the commencement and soon ceases, at any rate, for a time. 2. Those in which it persists. (1.) It has been pointed out by Dr. Radcliffe that this is the normal type of inflammation, and that the pain is to be referred to the preliminary stage of *irritation*, which is marked by coldness of skin, shivering, and general contraction of the blood-vessels; and there is little doubt that the nerves are in a condition of unusual feebleness, and of electric disturbance owing to that feebleness. But as soon as the second stage, or that of inflammation proper, accompanied at first with increased circulation, commences, spontaneous pain ceases, and only tenderness on pressure remains. (2.) But in certain cases pain persists after the congestive stage has commenced and the skin is hot and the pulse throbbing. These cases, however, are universally such as are distinguished by the fact that the tissues of the inflamed part are subjected either to compression, or to friction or stretching, both of which are only varieties of compression. The pain of subfascial suppuration, of pleurisy and peritonitis, and of gout, respectively, may serve as illustrations; and we cannot help being reminded of the tenderness to pressure which, in the ordinary course of inflammation, follows the cessation of spontaneous pain. It appears highly probable that the continuance of the pain during the congestive stage is the result of the accidental presence of pressure, more especially as the pain is severest in those instances (*e.g.* serous inflammations) where constant friction is going on.

Inflammation, however, is by no means, on the whole, an *elevation* of vital force; and though its first stage may be accompanied with increased activity of the circulation in the affected part, this does not continue to be the case; for the adhesiveness of the blood-corpuscles soon causes an obstruction, and at last an absolute stagnation of the flow. And correspondently with this, we find that the pain, though in an altered shape, often does return at a later stage with great severity.

It seems hardly necessary to say that pain produced by cutting or by bruising wounds is not due to stimulation; for in the one case the nerves of the part, and the tissues round them, are drained of blood, and in the other they are pressed on by the products of extravasation. Moreover, there is little doubt that cutting or bruising wounds diminish the normal electricity of the parts, another source of vital depression. Then as to the phenomena of the mortification of parts, it is well known that the pain is severest shortly before the nerves (usually the last structures to die) are about to become altogether dead. Again, in recovery from frost-bite, and from profound alcoholic intoxication, the nerves, during their struggle from temporary death, through low degrees of vitality, up to a state of healthy energy, are excessively painful.

Sir William Hamilton held the opinion that* "pleasure is the result of certain harmonious relations, of certain agreements; pain, on the contrary, the effect of certain inharmonious relations, of certain disagreements. Pleasure is a reflex of the spontaneous and unimpeded exertion of a power of whose energy we are conscious; pain a reflex of the overstrained or repressed exertion of such a power." The varieties of pain which have been already mentioned would appear to come under the latter head—"pains which are the reflex of *impeded* energy." With regard to the supposed class of pains which are "the reflex of an overstimulated energy," I would submit that there is some confusion of ideas here. The pain produced is not contemporaneous with the excessive energy, but with the exhausted condition in which the organ remains after the energy has operated. The brain, for instance, which has been laboring beyond its strength, has really been consuming its tissue faster than nutrition could repair it, until at last it came to have an imperfect energy, owing to the

* Lectures on Metaphysics, vol. ii, p. 440.

want of material in itself; and the result is a condition of the
nervous system which, as far as its outward symptoms go,
closely resembles the irritative stage of an inflammatory disease, such as I have endeavored to describe it, with pale,
cold, shrunken skin, and a tendency to shiver; it is under
such circumstances as these that the headache of an overworked brain sets in. The muscles of the boy who is going
through his first fencing-lesson have already been much exhausted before they got into that condition when the nerves
would convey the impression of pain to the sensorium. The
pain occurs during a state of *disorganization*, when it would
certainly seem that the energy must be *impeded*. And this
probability is increased on consideration of the fact that in
certain exceptional cases, when nutrition can be proportion-
ately increased to meet the excessive waste of tissue, the
sense of fatigue, and the pain which naturally accompanies
it, are not produced, or are produced only in a much less
degree than they would otherwise be.

An apparent instance of pain occurring as the reflex of an
overstrained energy, is to be found in the painful fatigue
which affects organs of special sense, such as the eye or ear,
when they have strenuously been directed, for long periods
of time, to the appreciation of particular objects. But the
supposition is doubly inappropriate in this case. It is inappropriate for the same reason as in the cases already alluded
to, because the painful feeling is not the accompaniment of
the energy, but of the exhaustion which succeeds it. And it
is further inappropriate, because it is irreconcilable with what
we know of the structure and functions of the organs of distinctive sense. For these organs consist essentially of nervous
expansions on the surface of the body, which passively receive
certain impressions (for the reception of which they are
specially fitted) of which the mind takes active apprehension.
When by an exercise of volition we place the organ in the
best position for receiving external impressions, and by the
exercise of attention provide that the mind shall take note of

these impressions, it will be found that the organ is capable
of receiving a certain number of them, for a certain time,
without fatigue, that is, without physical injury. But beyond
this limit it is impossible to go; the only result of any effort
in this direction will be a confusion of all the impressions received, and a sense of painful fatigue which is obviously the
reflex of *impeded energy* (using that word, energy, as Hamilton does, to express passive as well as active processes). The
receptive capacity **of an** organ of distinctive sense must
surely be proportioned strictly to the state of its nutrition,
and hence we can understand how that capacity may be im-
proved by anything (*e.g.* gentle and regular exercise) which
may improve the nutrition of the organ. Such improvement,
however, must be the work of time; meanwhile it appears to
be a contradiction in terms to say that an organ, whose duty
is passively to receive certain impressions, can be "stimu-
lated," at any particular moment, to receive more of those
impressions than its actual capacity is adapted for. With
regard to cases in which strong emotion seems to intensify
the acuteness of any sense to a preternatural degree, while at
the same time a **feeling of pain** is produced, it may be remarked that such cases arrange themselves under the head
of *special and exaggerated attention.* I ought, perhaps, to
apologize for venturing to quote the authority of a romance-
writer on a question of scientific observation; but Sir Walter
Scott was so accurate an observer of Nature (and of course
so unprejudiced as to this matter) that I am tempted to cite
his description of the sensations of Henry Morton when momentarily expecting death at the hands of the fanatic Cove-
nanters, in illustration of pain from exaggerated attention.
"His destined executioners, as he gazed around them, seemed
to alter their forms **and features,** like specters in a feverish
dream; their figures became larger, and their faces more disturbed; and as an excited imagination predominated over the
realities which his eyes received, he could have thought him-
self surrounded rather by a band of demons than of human

beings; the walls seemed to drop blood, and the light tick of the clock thrilled on his ear with such loud, painful distinctness, as if each sound were the prick of a bodkin inflicted on the naked nerve of the organ." Here we have an admirable description of the overmastering influence which emotion may exercise upon the finer and more distinctive sensations. In such circumstances as are here hypothetically represented (and also, as Dr. Chambers justly remarks,* in the disturbance of the nervous system, which often happens in fevers and in mania), some overpowering influence rivets the attention on certain objects, and correspondingly impairs the discriminative power of the organ of sense, so that false images of them are represented to the mind. To speak of such phenomena as proofs of a genuine hyperæsthesia (if by αἰσθησις we imply anything discriminative) is entirely incorrect. Sir William Hamilton has indicated the source of fallacy which is latent in all such statements,† viz. the neglect of clear distinction between perception proper and sensation proper. The former faculty takes cognizance of external objects in the way of *accurate distinction;* the latter merely appreciates the pleasurable or painful nature of the impression. The law of correlation between them is thus stated by Hamilton: "Sensation proper and perception proper must always co-exist; but so far from existing in equal degrees, their respective amounts are in exactly inverse proportion to each other. The more receptive of pleasurable or painful impressions an organ may be, the less adapted is it to the purposes of distinctive perception." In the section of the present volume which details my special researches on Anæsthetics some facts will be found which appear strongly to support this proposition. At present I can only state my general assent to its truth.

III. *Muscular Phenomena.*—The propriety of ascribing excessive or irregular muscular contraction to the influence of stimuli, exalting the nerve-force, cannot be discussed at length here; and considering how fully the subject has been

---

* Renewal of Life, p. 68.   † Notes on Reid, p. 886.

treated by Dr. Radcliffe in his work on Epilepsy, and in his recent lectures at the Royal College of Physicians, there is little necessity for me to go over the ground again. In the chapters on true Stimulation, and on the Phenomena of Narcosis, however, proof will be adduced that one of the most ordinary effects of stimuli is to check convulsive muscular movements; and, on the other hand, that such movements are part of the regular sequence of phenomena in the action of paralyzing doses of many narcotics, and are occasionally produced by every true narcotic. Relying on the accumulated testimony here indicated, I must content myself now with the simple statement, that the occurrence of irregular or convulsive muscular movements affords no proof whatever of any antecedent stimulant action on the organism.

IV. *Secretory Phenomena.*—The assumption that increased secretion necessarily implies stimulation of the secretory glands, or of the nerves governing them, has been conclusively proved to be unfounded, by the researches of Bernard and Brown-Séquard on the effects of section of the sympathetic nerve. This operation, when performed on the cervical trunk of the nerve, is followed by intense congestion of the vessels beyond the point of section, contraction of the pupil, and a copious secretion or flux from the several glands of the corresponding part, and similar results have been obtained from the section of the sympathetic branch going to the submaxillary gland. Dr. Gairdner and others have observed cases in which pressure on the sympathetic nerve at the root of the neck by tumors has occasioned flushing of the face and copious unilateral sweating. I am able to cite one remarkable case from my own practice, in which similar effects were, in all probability, produced by paralyzing pressure on the abdominal branches of the sympathetic.

Case.*—W. R., æt. five years and nine months. Was a healthy-looking child when born, but at three days old sud-

---

* By one of those mishaps that one can never sufficiently regret, some carelessness in the taking of this case renders the account of the state of the pupil unreliable; and I have, therefore, omitted all notice of this point.

denly had a series of fits, of which it is impossible to obtain any accurate description; remained insensible for three days, and then recovered, with no sign of anything amiss except an inward squint of the left eye. He continued in good health until he was about three years old, when he began to experience a constantly-increasing "choking sensation" in the throat, and it was observed that the temperature of the left cheek was higher, and its color deeper, than that of the other side; *moreover, the left side of the face was frequently covered with perspiration.* After this had lasted a few days, he suddenly became insensible, and foamed slightly at the mouth; the left arm and leg became quite rigid, and were repeatedly jerked toward the median line. This condition lasted for about half an hour. Since that time about four more fits had occurred (the last about seventeen months previously to my first seeing him); the same phenomena preceded and accompanied all of them. In the intervals there had always remained a certain amount of weakness of the whole left side, and a considerable tendency to rigidity of the left arm and leg; the act of putting on the left boot often caused the leg to be spasmodically extended. Recently the bowels had become constipated, and, as always happened when this was the case, the choking sensation had reappeared, and a fit appeared to be imminent. The boy was brought to me at the Chelsea Dispensary, October 31, 1862. He was a well-grown and well-nourished child, with a rather pale complexion, a considerable internal squint of the left eye, and a dragging of the left leg as he walked. The left side of the face and head was slightly hotter than the right, and there was a very perceptible flush on the left cheek. On handling the muscles of the left arm and leg, it was observed that they had a marked tendency to tonic contraction. The tongue was red and clean in the median line, over a space of about one-eighth of an inch on the left side of it, and over the whole of the right half. The outer three-fourths or thereabouts of the left half of the tongue was covered with a thick creamy fur, through which numerous bright red papillæ protruded. The bowels had not been opened for two or three days; the patient complained of nausea and a choking feeling in the throat. A very mild aperient mixture was ordered to be taken three times a day. This relieved the bowels regularly, and in the course of the next fortnight the threatening symptoms **had** almost entirely disappeared. He now **ventured** to omit the medicine, but in three days' time was brought to me in a much worse condition than when I first saw him. No motion of the bowels had taken place for forty-eight hours; the whole left side of the face and the left ear were flushed deep red; the surface was pungently hot, and *bedewed with a copious sweat*—the heat and sweating were found to extend over the whole left half of the head, and to cease abruptly at the median line. Before he could leave the room, after I had prescribed for him, he was seized with an epileptiform fit precisely similar to those which have been already described, and which lasted twenty minutes. On recovery, he seemed heavy and semi-comatose for an hour or two. On visiting **him** at the end of this time, he appeared much as usual, except that the unilateral flushing and perspiration and the feeling of choking were still present. A brisk purgative enema was now administered, and a large quantity of scybalous fæces were discharged: this was almost immediately followed by the entire disappearance of the unilateral heat and sweating and of the choking sensation. The boy recovered **his** ordinary appearance, and was desired to continue the mild aperient medicine, and also to take half an ounce of cod-**liver** oil per diem. From this time forward, the boy never had a bad symptom: he was retained under observation for two months, and then discharged quite well, except that he had still a slight weakness of voluntary power in the right arm and leg.

The paralyzing distention of the bowels by fæces, which seems to have produced such serious effects in this remarkable case, is exactly the same agency as often produces violent convulsive outbreaks in general paralytics.* There can be little doubt, I think, in the mind of any one who will impartially examine the facts, that the condition of the nervous apparatus is one of paralysis, in the one case and in the other. Henceforward there can be no sort of propriety in assuming that the flushing and sweating so commonly produced by an overfull or an ill-digested meal is an active phenomenon, depending on stimulation: indeed, the entire phenomena of excessive and untimely secretions from glands must be considered open to a possible interpretation, the reverse of that which represents them to be the effects of stimulant action. There is too much reason to believe that many medicines which are supposed to have specific effects in increasing the flow from particular glands do, in truth, only cause a passive flow of a fluid containing *some* of the normal elements of the secretion, in virtue of their paralyzing influence upon the nervous apparatus of secretion. Nevertheless, the name of "specific stimuli" has been unhesitatingly accorded to such medicines hitherto: a practice which is decidedly objectionable in view of the facts of modern physiology.

V. *Circulatory Phenomena.*—Increased rapidity of the heart's action is one of the most common items of evidence appealed to as indicating the operation of a stimulant upon the organism. This subject involves considerable difficulties; for there is little doubt that a moderate increase of the heart's action beyond its normal rapidity often does prove to be of real advantage to the vigor of the faculties, for the time, and is followed by no depressive reaction whatever. On the other hand, no fact is now more familiar to practitioners than that the most purely debilitating influences are precisely those which most greatly increase the rapidity of the pulse. No exact line of demarkation can probably be laid down: but we are entitled to believe that an increase of a few beats per minute, *when the strength of the heart's action is simultaneously maintained or increased*, is a healthy phenomenon, one

* *Vide* Austin, "On General Paralysis."

which is within the order of Nature, and so far entitled to be described as an effect of stimulation; but that, on the other hand, too great an increase of frequency *always implies debility*, more especially when coupled with diminution in force, or irregularity in rhythm.

As for the violent throbbings which often attend local inflammations, there is little reason for attributing them, as is often done, to the results of stimulation, or irritation. The advance of pathological inquiry is more and more rendering it probable that this phenomenon is of purely mechanical origin; being, in fact, due to the opposition with which the blood meets in its progress forward through the vessels of the inflamed part.

VI. Lastly, we have to consider the propriety of indiscriminately applying the term Stimulation, as is commonly done, to all processes which result in increased nutritive, or formative action, in a part of the body.

This question forms the very central point of the discussion as to the essential nature of inflammation. We have to consider whether such a change as that which is set up in a tissue, let us say, like cartilage, by drawing a thread through it (leaving it imbedded as a foreign body), deserves the name of stimulation.

What are the changes produced by such a lesion, in such a tissue as cartilage? In the immediate neighborhood of the damaged part all the cells are enlarged, indicating a distinctly increased nutrition of them. If now we inquire what is the result of a higher degree of irritation, such as that which is present in the case of so-called "ulceration" of articular cartilage, we find that a greatly increased *proliferation* of the cells is the characteristic feature of the process in parts which are strongly affected, the formative activity being indicated by the presence of mother-cells which are crowded with numerous contained cells, probably representing successive generations, or more often with nucleoli. And in the parts which actually border on the focus of irritation, we perceive unmistakable evidence that the exuberant formative process is attended with contemporaneous *degeneration*. The inference would seem plain that the agency which produces all this mischief is a debilitating, devitalizing one; and that all this ill-timed and hasty excess of nutrition and formation is but the result of a shock to the vital condition of the part, of which the highest expression is seen in the positive degradation of the new-formed matter, in the most strongly affected localities, from the very moment of its generation. We are not justified in speaking of such a process as this (which is, in truth, a type of all inflammatory processes) as an effect of stimulation.

So again with the remarkable phenomena which are so often produced by diseases which injure the vitality of nerves which are distributed in the neighborhood of epithelial surfaces. It is by no means uncommon, in cases where the fifth nerve of one side is affected with severe neuralgia, and is obviously in a state of great depression, to find the tongue coated, on the side toward the diseased nerve, with exuberant epithelium, and to observe that the thickness of this covering varies in direct ratio with the severity of the pain and the intensity of the depression. In the curious case above quoted of epileptiform convulsions dependent on loaded bowels, the same effect appeared to be produced through the sympathetic nerve of the affected side. And in the different varieties of herpes, there is very strong reason to suspect that the cause of mischief lies in the depressed vitality of nerves which supply the skin in which the inflammation occurs. In all these cases the phenomena are essentially the same in their nature and order of occurrence: first of all, depression, evidenced both by the general sensations and by the occurrence of local pain; and then, when this depression has lasted a certain time, and reached a certain depth, a hasty and imperfect, though multitudinous formation of cells and other elements of tissue, with a positive tendency to decay even in the very moment of their birth.

Activity of reproduction is not to be taken as the standard of life, either in the organism or in any part of it. If, indeed, we knew that every action in organized beings, which to our eyes looks different from what the inorganic world can exhibit, was really a peculiar attribute of "life," it would then seem to follow that the excess even of so humble a process as that of the mere multiplication of cells and fibers was a proof of superabundant vitality. Far different, as it seems to me, is the lesson inculcated by the actual state of our knowledge of physical laws; a lesson which I shall do my best to give brief expression to hereafter. We have no right to make the assumption in question, nor to use the test which is based upon it.

If excess of formative action be rejected as a criterion of vital stimulation, the same course must be adopted with regard to *excessive nutrition*. Doubtless, nutrition is a necessary condition of life; the functions of no organ can be performed unless its elementary tissues receive a certain amount of development, and maintain a certain standard of bulk amidst the changes to which they are subjected. But it is by no means logical to infer from this that the amount of vitality is proportionate to the bulk of tissue, whatever this may be. Nor is it reasonable, because life implies constant change, to assume that the amount of change is the gauge of life. And as nutrition can only minister either to mere bulk or mere rapidity of change (when it passes a certain necessary line), we are entitled to reject all excessive and useless nutrition from the list of phenomena which prove the fact of antecedent "stimulation."

CHAPTER III.

THE DOCTRINE OF STIMULUS.—SUGGESTIONS FOR ITS
RECONSTRUCTION.

HAVING now reviewed the history of the rise and progress
of the doctrine of Stimulus, and having indicated some of the
most striking points of discrepancy between the popular views
on this subject and the teachings of modern physiology, we
have to inquire whether some new and more rational interpretation cannot be found for a word which it would be
highly inconvenient to dismiss from use.

In the first place, it seems obvious that etymology should
be respected in any new application of the word which we
may resolve to make. We should not forget that Stimulus
means a *spur* or *goad*. On the other hand, we must learn
to think of the expression as figurative only; once having
realized this, we shall be the less likely to apply it to physiological processes with which it has no connection. It has
been already mentioned that John Hunter defined stimuli as
agents which increase some natural action or tendency as
opposed to irritants which produce actions which are altogether abnormal; and this, if we reflect on it, will appear to
be the most reasonable application of the term. We must
take our ideas of stimulation, not from the notion of a sharp
thorn plunged suddenly into a sensitive tissue, but from that
of a goad applied to the sluggish ox, whose sensibilities it
wounds no further than is sufficient to remind him of work
which he has to do, and can do very well if he likes to try.
And even this is but a figurative and a very imperfect account of therapeutical stimulation, for the latter not improbably includes, as I shall endeavor to show, not merely

8

the impulse to act, but a supply of the materials which are necessary for action.

It is impossible for us to advance a single step toward a satisfactory explanation of the true office of material stimuli, until we have determined what it is to which we suppose these physiological agents to be applied; whether we are to consider them as operating upon physical objects, under the guidance of the laws of the material universe, or as operating upon something resident in the organized body, which is *not* exactly physical. Our first duty, therefore, must be to state as clearly as possible the meaning which is to be assigned to the word Life, as far as it relates to corporeal processes.

In the opening chapter of the present section it was shown that there is a true historic continuity of vitalistic ideas, from the earliest quite down to the present time. In the present day, however (pneumatology having gone out of favor with metaphysicians), we hear little or nothing of Vital Spirits.* In conformity with the general tendency to give dynamical explanations of all physical processes, it has become the fashion to assume the existence of a "vital force," similar in its nature to heat, light, electricity, and other known physical forces, but peculiar to organized beings; and it is this vital force which is supposed to be generated, or educed, in some way which it is not very easy to understand, by the action of stimuli.

This view is very strenuously supported by Dr. Inman, who certainly has not dealt with the question carelessly, and who is a thoroughly original thinker. "The words *vital force*," says this author,† "signify the living principle which exists in every organized being. It is the power which enables the eye to see, which a camera obscura cannot do; which converts inanimate matter into combinations that the

---

* The last systematic attempt to restore such existences to popular favor was, I believe, the Bridgewater Treatise of Prout "On Chemistry, Meteorology, and the Function of Digestion, considered with reference to Natural Theology." *Vide* p. 309, Bohn's edition.

† Foundation for a New Theory of Medicine, 2d edit. p. 13. Churchill.

chemist cannot imitate, which frames out of the same materials various organs, each having a different function. . . .
It is not simply the result of a fortuitous concourse of atoms,
for they may be present, and no life be there. It is not
nerve force, for that is dependent upon life, and a dead nerve
is as powerless as a stone. It is not organization alone, for
organs may be present, as in a corpse, where there is no life
or vital force." . . . It is "a force completely *sui generis,*
but one which, being more or less allied to physical forces,
has its phenomena more or less modified thereby." Dr. Inman supports this position by numerous arguments derived
from the striking differences exhibited by members of the
same organic family, whose structure, nevertheless, may be
built up, apparently, of exactly the same physical elements,
and subjected to exactly the same physical influences; from
the extremely different effects of different organic matters
when introduced in the system of a healthy individual (*e.g.*
variolous and leprous serum); the fact being, that each kind
of organic matter thus introduced has a vital principle of its
own, which, and which only, it can transmit, &c.

Without attempting to answer Dr. Inman's arguments in
detail, I would call attention to a remarkable development of
recent scientific thought, the first traces of which may be
seen in the writings of the greatest thinker of the present
century. Arguing against the artificial separation commonly
made between physical and vital forces, Coleridge says:*
"To a reflecting mind, indeed, the very fact that the powers
peculiar to life in living animals *include* cohesion, elasticity,
&c. or (in the words of a late publication) 'that living matter
exhibits these physical properties,' would demonstrate that
in the truth of things they are homogeneous, and that both
the classes are but degrees and different dignities of one and
the same tendency. . . . The lower powers are *assimilated,*
not merely *employed,* and assimilation presupposes the homogeneous nature of the thing assimilated; else it is a miracle,

* The Theory of Life, p. 42. Edited by Seth B. Watson, M D. &c. Churchill, 1845.

only not the same as that of a *creation*, because it would imply that additional and equal miracle of annihilation. In short, all the impossibilities which the acutest of the reformed divines have detected in the hypothesis of transubstantiation would apply, *totidem verbis et syllabis*, to that of assimilation, if the objects and the agents were really heterogeneous. Unless, therefore, a thing can exhibit properties which do not belong to it, the very admission that living matter exhibits *physical* properties includes the further admission that those physical or dead properties are themselves vital in essence, really *distinct*, but in appearance only *different ;* or in absolute contrast with each other."

The position maintained by Coleridge, that the so-called inorganic world is as truly, though in a lower way, informed with "life," as is the world of organized beings, was doubtless a startling one ; and it is a striking proof of the foresight and penetration of that great man, that our latest scientific inquiries appear to be tending toward a confirmation of his views. The remarkable researches of Mr. Grove* have established the fact "that the various imponderable agencies, or the affections of matter which constitute the main object of experimental physics, namely, light, heat, electricity, magnetism, chemical affinity, and motion, are all correlative, or have reciprocal dependence. That neither, taken abstractedly, can be said to be the proximate or essential cause of the others, but that either may, as a force, produce or be convertible into the others; thus heat may mediately or immediately produce electricity, electricity may produce heat, and so on of the rest." And Mr. Grove expresses his belief "that the same principles and mode of reasoning as have been adopted" in his essay "might be applied to the organic as well as to the inorganic world ; and that muscular force, animal and vegetable heat, &c., might, and at some time will be, found to have definite correlations."† Dr. Carpenter goes

---

* Correlation of the Physical Forces, 4th edit. Longman, 1862.

† "On the Mutual Relations of the Vital and Physical Forces." Philosoph. Transactions, 1850.

a step further, and arrives at the conclusion, that not merely are the *materials* withdrawn from the inorganic world by vital agencies, given back to it again by the disintegration of the living structures of which they have formed a part, but all the *forces* which are in operation in producing the phenomena of life are in the first place derived from the inorganic universe, and are finally restored to it again. The arguments of Dr. Radcliffe,* which were referred to in the first chapter as affording a startling proof of the unsound basis on which the popular theory of stimulus rests, regard the subject of vital forces from a point of view which annihilates any fundamental distinction between them and the forces of inorganic nature, and assert a complete unity between living organisms and the rest of the universe, both as to *form* and as to *force*.

It is impossible here to recapitulate the several arguments by which the authors above quoted support their conclusions; but the mere fact that such a chain of reasoning has been wrought out under the influence of modern physiological research, by men of such mark, is itself a weighty reason for suspecting that truth must at least lie somewhere in the directions to which their speculations tend. Under these circumstances, it would seem to need something more than the arguments, chiefly negative, which Dr. Inman so ingeniously propounds, to command our assent to the proposition that vital force is a principle completely *sui generis;* and the words of Mr. Lewes† appear to be justifiable. "A vital principle," says Mr. Lewes, "is incapable of proof. If it exist we cannot know it, and unless its existence can be proved, it is to us a mere phrase concealing our ignorance." We are not called upon to *account* for Life, but only, as far as may be, to *explain* it. "The former process," as Coleridge forcibly remarks, "would imply the statement of something prior (if not in time, yet in the order of nature) to the thing accounted for, as the ground or cause of that thing. . . .

---

* Proteus, or the Law of Nature. Churchill, 1850. Epilepsy and Convulsive Diseases. Churchill, 3d edit. 1861. Vital Motion. Churchill, 1851.

† Physiology of Common Life.

And to this, in the question of Life, we know no possible answer, but God. To account for a thing is to see into the principle of its possibility, and from that principle to evolve its being. Thus the mathematician demonstrates the truths of geometry by constructing them. . . . To explain a power, on the other hand, is (the power being assumed, not comprehended. . . .) to unfold or spread it out: *ex implicito planum facere*. In the present instance, such an explanation would consist in the reduction of the idea of life to its simplest and most comprehensive form or mode of action; that is, to some characteristic *instinct* or *tendency*, evident in all its manifestations, and involved in the idea itself." Or to speak in other terms, the explanation would be like that of an inquirer, who, presuming to know nothing of the *power* that moves the machine which he is investigating, takes those parts which are presented to his view, seeks to reduce the various movements to as few and simple laws of motion as possible, and out of their separate and conjoint action proceeds to explain and appropriate the structure and relative positions of the works. " In obedience to the canon, ' Principia non esse multiplicanda præter summam necessitatem cui suffragamur non ideo qui causalem in mundo unitatem vel ratione vel experientiâ perspicimus, sed illam ipsam indagamus impulsu intellectûs, quia tantumdem sibi in explicatione phænomenorum profecisse videtur quantum ab eodem principio ad plurima rationata descendere ipsi concessum est.' "

I cannot imagine that there can be any satisfactory investigation of the subject on which we are engaged, or of any similar one, apart from an observance of these canons of inquiry. It is enough for us to know, by evidence which is overwhelming, and on authority which we cannot dispute, that there is one part of our nature, the immortal soul, which is altogether mysterious, and of the true laws of which mere human science can tell us very little. There is nothing for it, in this case, but to acknowledge the impotence of observation and reason; but surely we show little wisdom in gratuitously assuming the existence of a further insoluble mystery, in the shape of a vital principle.   And when we go beyond this, and proceed to endow this mysterious and incomprehensible principle with properties which are the faint and ghostly reflection of the characteristics of physical forces with which we have some acquaintance, I cannot but think that we commit an error of the very gravest kind.   Pathology and therapeutics have been surging backward and forward for many a long day, at this very point.   So long as we persist in assuming that the effects of medication are mysterious to us, not because we have not toiled long enough at physical science to understand them, but because they are produced through the agency of forces which are neither wholly physical nor wholly spiritual, so long shall we be incapable of real progress.

Let us therefore be content to "reduce the idea of Life to its simplest and most comprehensive form or mode of action; that is, to some characteristic instinct or tendency, evident in all its manifestations, and involved in the idea itself;" and we shall agree, I think, with Coleridge, that the only such tendency to which we can ultimately refer all the phenomena, is that of—Individuation, "the internal copula of bodies . . . the power which discloses itself from within as a principle of *unity* in the *many*."   But if this be the utmost limit to which we can go in the way of definition, it is obvious that we have no right to assume that the increase of any one or more of the "many" forces concerned in the life of the body, is an exaltation of the "vitality" of the organism, or of the part wherein such increase of force is developed.   The standard of life is a certain exact balance of various forces, developed with a certain *constant relation to* material tissue arranged *in a definite manner;* to say that we increase such life or "vitality" in one part of the organism by destroying this balance, is a contradiction in terms.   And the standard of function in an organ is the accurate discharge of such an amount and kind of work as may help to maintain this healthy adjustment of power and of matter in the organism :

to say, then, that an organ exhibits an increased activity merely because it is seen to be under the influence of extraordinary powers, and to present a new arrangement of matter, is incorrect. It follows that the use of the word "stimulus" to express the cause of *excessive* action of any kind in the body, is improper, and we may expect to find that when a remedy really does improve vital power it does so by restoring, in a natural manner, the natural condition of things. Let us accept the facts that we can gather for ourselves, and see whether this be not the real action of stimulants.

Genuine Effects of Stimulation.

When these are carefully separated from apparent but unreal instances of this kind of physiological action, such as were discussed in the preceding chapter, they are divisible into cases of—

    I. Relief of pain.

    II. Removal of muscular spasm, tremor, or convulsion.

    III. Reduction of undue frequency of the circulation.

    IV. Reduction of excessive secretion.

    V. Removal of general debility, or of special fatigue of muscles, brain, or digestive organs.

    VI. Removal of delirium, or maniacal excitement, and production of healthy sleep.

    VII. Support of the organism in the absence of ordinary food.

VIII. Local increase of nutrition *where this is deficient.*

The merit of perceiving that most of the above effects can be produced by stimulation better than in any other way, certainly belongs to Brown; and there is hardly any more striking instance, in medical history, of important observations undeservedly neglected, than is afforded by the fate of the "Elements of Medicine." The theory contained in that work was, doubtless, justly condemned; but many of the facts recorded are of precious significance, and should have been carefully remembered. I shall have frequent occasion to refer to them in the remarks which are now to follow.

I. The power to relieve pain is a property which is commonly supposed to be pre-eminently characteristic of sedative, depressing remedies; but nothing can be more unjust than such an idea. Such remedies will indeed relieve pain, but only by a process of *poisoning*, which puts life in more or less danger, and which will be described later as "True Narcotism." The mode of treatment which, ninety-nine times out of a hundred, proves successful in removing pain in medical practice, is pure stimulation. As examples of the remedies which act in this way, we may mention—

1. Quickly digested and nutritious food. 2. Opium, in doses of one to two grains; or morphia, in doses of a quarter to half a grain. 3. Carbonate and muriate of ammonia, in doses of five and ten grains respectively. 4. Alcohol, in doses just too small to produce flushing of the face, or sweating of the brow. 5. Chloroform, inhaled (in the proportion of about two per cent. to the bulk of atmospheric air) for a short time; or taken internally, in doses of a few drops. 6. Certain fetid gum-resins. 7. Many aromatic volatile oils. 8. The bitters, pure and aromatic. 9. "Counter-irritation," as it is called; stimulation, as it should be termed, through the adjacent skin with mustard, turpentine, &c., or with blisters. These are some of the more rapidly-acting remedies.

At the very head of this list it will be seen that I have placed food. I do so because if food can be digested there is nothing which in so many cases relieves pain with satisfactory celerity, unless it be kept up by a mechanical cause. The action of all the other remedies in the list is a pale reflex of that of highly nutrient and easily-digested food. The same increased firmness, without hurry, of the pulse; the same renewed cheerfulness, less of the sense of fatigue, feeling of grateful warmth throughout the body, in each case accompany the relief of pain, in a greater or less degree. Brown states, that the headache and pains in the joints which attend asthenic diseases (and asthenic diseases with Brown formed 97 per cent. of all maladies) are easily curable by stimulant plans of treatment (among which he includes the use of ordinary food). From personal experience, I can speak as to the effects of food in relieving the pains of two very different affections—the first of a neuralgic kind, and the second from intense cellular inflammation of the finger and hand owing to a dissection wound. And I have seen one patient suffering severe agony from peritonitis, who derived rapid relief from the careful and gradual injection of a pint of rich soup into the rectum.

These statements will excite a certain amount of surprise, perhaps even of incredulity; for they are certainly in direct opposition to the teaching which most practitioners must have had impressed on their minds. That common food should exercise a salutary influence both in neuralgic and in inflammatory pains is what we certainly were not taught, in our class-books, to believe or expect: but I submit that inveterate preposession on this head has prevented anything like a fair trial of the question. In inflammatory affections attended with acute pain, there is often another reason for not trying such a plan of treatment, viz.: the inability of the stomach to retain any considerable quantity of food. In such a case the plan of injection into the rectum might be tried, and with good prospect of success, because in most of these cases the system is suffering severely from a compelled abstinence from food.

Case.—The case above referred to, of peritonitic pain relieved by the administration of food, was that of a young girl, aged fourteen, in whom there was every reason to believe that a group of enlarged mesenteric glands, probably tuberculous, had suppurated; possibly some pus had suddenly been discharged into the peritoneal cavity; at any rate the excessively rapid and small pulse, the posture of the patient, the expression of countenance, and the intense and persistent vomiting, which caused everything swallowed to be immediately rejected, left no doubt, when taken together with the character of the pain, of the general inflammatory affection of the peritoneum. Leeches, turpentine stupes, and opium in large doses had been tried before I saw the case, but utterly in vain: the vomiting was worse than ever, the pain excruciating. I ascertained that decided, though temporary relief, had been obtained at one time after the use of a warm gruel enema, which however had produced no evacuation of fœces. Suspecting that the enema might have given relief by reason of part of it having been absorbed and having acted as nourishment, I ordered a **pint of** strong **meat soup to be slowly** thrown, in three successive portions, into the rectum. **Within a quarter of** an hour a most sensible relief of the pain was felt, and the expression of anxiety and distress in the countenance had almost disappeared: the amendment thus **produced** lasted for about three hours, when the pain again returned, though not with its original violence. A new injection of soup was followed by the same relief to the pain as previously: moreover, the pulse fell in about an hour's time to 104 (at the time of the first visit it was 124); the skin became **much** cooler and slightly moist; the tongue assumed a more healthy appearance, and the stomach so far lost its irritability as to bear the presence of small quantities of broth and wine, which were now ordered to be administered every two hours. In forty-eight hours from the time of the first injection, the pain had altogether disappeared. **All** action of the bowels was carefully prevented for several days, **and the patient** made a complete though tedious recovery.

Of all stimulant medicines there is none which approaches **more closely to food in its manner of relieving pain than** opium administered in small doses. Brown indeed considered that this drug was far more potent as a stimulus than **any** other substance; but this is certainly a mistake, for nothing acts so rapidly as ordinary food in cases where **it is suitable.** It ought, moreover, to be clearly understood, that there are two ways in which opium may relieve pain, viz.: by its employment as a stimulant, in small doses, and its use as a true narcotic, in larger quantities. It is with the former alone that we have to do at present. Dr. Inman has called attention to this kind of action of opium, and has justly observed that it can only take place in consequence of the drug exerting a modifying influence on the organic condition of the affected part; in fact, that opium administered in this way **averts** a threatened destruction of tissue, and consequently relieves pain. This idea is perfectly consonant with the opinions as to the physiology of pain which were expressed in the preceding chapter. It must be allowed that opium is a drug which more than any other is liable to produce varying effects, according to the idiosyncrasy of the patient to whom it is administered: nevertheless, the careful collation of a large number of observations on its action in small doses (relatively to the idiosyncrasy of the taker) has convinced me that in this sort of use of the drug we produce no narcotic or depressing effects whatever. My own personal experience on this point consists of two parts—observations on the action of one-grain and two-grain doses of opium, or an equivalent quantity of a salt of morphia, taken during a state of full health, and of the same taken for the relief of depressing catarrh of an influenzal type accompanied with severe pains in the limbs. One grain of opium, or one-fourth of a grain of morphia, taken by me when in a state of health, produces scarcely any appreciable effect, supposing that I am in a state of indifference as to food (*i.e.* with a stomach neither recently filled nor, on the contrary, pained with hunger). Administered at a time when I am decidedly hungry, it has a most remarkable influence in dispelling the desire for food, for which it substitutes a warm and comfortable feeling, not merely in the stomach but also in the whole body, which exactly resembles the pleasant sensation occasioned by a moderate meal of good food, and endures for about an hour or an hour and a half, and then quickly subsides, leaving no trace of depression except that caused by hunger. I have repeated this experiment at least a dozen times, at intervals of several weeks; and always with the same result (except that on one or two occasions the influence of the drug had lasted longer). The medicine caused no particular inclination for sleep. The largest dose of opium which I have been able to take without producing greater effects than those mentioned is two grains, of morphia half a grain; any increase beyond this amount has invariably produced some of those symptoms which will be described in the section on "Narcotism." The other way in which I have personally tested the effects of opium is as a remedy for influenza-like catarrh, of which I rarely escape one or two attacks every winter. These attacks are very sudden, and are attended with great depression, with very severe aching pains in the head, back, and limbs, and much febrile disturbance. **For this sort of** ailment opium, if taken early enough, is a perfect remedy. One-third of a grain of muriate of morphia is taken at bedtime: its influence begins to be felt, in from twenty minutes to half an hour, in the gradual disappearance of the sense of intense weakness, the relief of the pains, and that peculiar feeling of thorough and evenly distributed warmth of the whole body, which is so different from fever on the one hand, or chilliness on the other. Natural sleep at last supervenes, and the morning finds me ready to wake at the usual time, with a moist and clean tongue, a total absence of pains and of fever, and an excellent appetite; but little food, probably, having been taken on the previous day. The effects of an overdose of morphia, taken with the mistaken notion of making sure of a cure, will be described further on.

That the pains of severe catarrh are removed by the above-mentioned doses of morphia acting in a stimulant manner, is, **I think,** proved, first by the fact that the same result can be obtained without the aid of this medicine, if the strong repugnance which I usually feel at these times to food and wine can be got over and a fair quantity of either be taken; and secondly, that there is never any, even the slightest, depressive collapse as an after-consequence. Out of forty patients in whom the same treatment was adopted for the relief of similar catarrhal pains, only six failed to experience considerable relief: in all these six cases the opium disagreed more or less, causing **sickness,** constipation, and headache, and rather aggravating the patient's distress ; but all those by whom the peculiar feelings of relief which have been described were experienced were exempt from headache or after-depression **of any sort.** The use of opium for the relief of severe catarrh, **is a very** old and well-known treatment, and is satisfactorily justified by accurate clinical observation : but it is important that we should **not forget that** the effect we desire may be obtained, if at all, by small doses ; small, that is to say, in **proportion** to the patient's idiosyncrasy and habits. The use of full narcotic doses is both needless, and, in most cases, injurious.

The pain of inflammatory affections is remedied, so far as it can be *directly* remedied, by the stimulant action of opium, and the same moderate dose only is required to produce the effect: but this must be repeated more frequently than would be necessary in a more trifling affection, probably because the reparative power of the system is more deeply disordered. Thus the dose of morphia which is sufficient in the average run of cases to relieve the pain of acute pleurisy is one-third of a grain; but it needs to be repeated about **every** third hour. Again, **in** peritonitis, if opium will relieve at all, it will relieve in a moderate dose, except in those cases where the **patient** cannot keep his muscles relaxed and quiet, or in which the peristaltic action of the bowels goes on. It is by paralyzing the movement of the intestines, or by paralyzing the brain to the perception of pain, that larger doses sometimes appear to relieve when smaller ones have failed.

It need hardly be remarked, that the pain which occurs in a distant organ sympathetically (to use a common phrase) with disorder of the digestive organs, is not to be relieved by opium in any shape: **this** is not because of any radical difference in the pain itself from that of pure neuralgia (which notoriously *may* be temporarily relieved by such a dose as a grain or two grains of opium), but because the opium in this case is not quietly digested, and consequently aggravates the mischief on which the pain depends.

Such, briefly, is the action of opium in relieving pain, when it is used in judicious medicinal doses: let us now compare it with that of ammonia, in the forms of carbonate, sesquicarbonate, and muriate. The carbonate forms the basis of that favorite remedy, *sal volatile;* but as there is alcohol in this preparation, it is perhaps better not to talk about the influence of the ammonia as a thing to be distinctly known. The sesquicarbonate, however, dissolved in simple water, has an action which is obvious, and not to be mistaken. After making comparative experiments on myself with grain doses of opium, and five-grain doses of sesquicarbonate of ammonia, I am prepared to affirm that the action of the two substances (in these respective doses), in health, is as nearly identical in its general result as it is possible to imagine them to be, with the exception of differences in taste, and in the rapidity of diffusion. And from experiments, both on myself and others, it has become obvious to me that they are about equal in their power of relieving pain, and that, so far as can be judged from outward signs, they relieve it in the same way. Bating such slight differences as have been named, the description above given of the way in which opium relieves pain in small doses may apply in so many words to the effects of sesquicarbonate of ammonia in the dose of from three to five grains. The muriate (sal ammoniac) is a medicine which I suspect is less generally used, and its powers less highly rated than they should be. In doses of from ten up to twenty grains (more than this is a needlessly, and I think perhaps mischievously, large dose), this drug has a pure and powerful stimulant action, which lasts for a considerable time, and among its other effects frequently relieves any pain, more especially of the neuralgic kind. I have known severe neuralgia, when not of long standing, removed permanently by a single dose. Both the carbonate and the muriate produce effects some of which last for several hours, although the more obvious disappear more quickly: the practical advantage in using the muriate consists in the fact that a larger dose may be given without irritating the stomach. No depressive reaction follows the use of a moderate dose of either salt.

There is no adequate reason for separating the action of the carbonate and muriate of ammonia from that of opium administered in small doses, such as I have mentioned, so far as their effects on pain are concerned. Even in inflammatory affections, such as pleurisy, I am strongly inclined to believe that the salts of ammonia have a direct influence in relieving pain. It is customary to ascribe to their diaphoretic action the diminution of pain which often follows their use in serous inflammations: but this notion is of the vaguest, and rests on no evidence that will bear examination; although the sweating may be, and probably is, beneficial to the patient. Why it is that for pain in general the carbonate and muriate are the more frequently serviceable remedies, while for the relief of the particular pain which accompanies serous inflammations the citrate and acetate are more valuable, I am not prepared to say; but the fact does not depend on any difference in their diaphoretic power, if indeed any exist, which is doubtful.

One of the most powerful remedies which can be used for the relief of pain is Alcohol. In cases of neuralgia there is sometimes no medicine whatever which will produce so great an effect in relieving the sufferings of the patient as a full stimulant dose of alcohol. There is no necessity to produce any of the narcotic or intoxicating effects: we may stop short of this point with distinct advantage: by carrying the treatment any further we not only expose the patient to the certainty of the unpleasant depression which accompanies the recovery from alcoholic poisoning, but there is a considerable danger of the neuralgic pain recurring with as much violence as ever during the period of that depression. An illustration of this was afforded by a case of severe facial *tic* which came under my care, and which will be referred to presently under the head of "Treatment with Food-medicines." The patient told me that occasionally, when the pain was more unbearable than usual, he resorted to brandy-and-water for relief, and always obtained it, for the time. He told me also that a small dose (according to his habits when in health) was sufficient to relieve the pain in about half an hour from the time of taking it; but that sometimes, when the agony was especially acute, he had been tempted to hasten the relief by taking two or three glasses instead of one; under these circumstances the pain ceased, but a heavy comatose sleep supervened, from which he awoke shivering and depressed and in such a state that the slightest puff of cold air, or even a sudden movement, would cause a new access of neuralgic pain. On the other hand, a moderate dose would usually completely relieve the pain, which might not recur for days.

The treatment of pain by *counter-irritation* unquestionably produces its beneficial effects by stimulating the circulation of the painful part when this has been unnaturally depressed. If it were in our power to put the blister, or turpentine stupe, actually upon the seat of pain, we should so far destroy the vitality of the tissues that the pain would be aggravated. As it is, we apply our "counter-irritant" at such a distance from the painful part that we produce a healthy quickening of circulation in it, while very probably at the same time electric changes are effected in the nerves, which aid in the beneficial effect.

If now we turn our consideration to the remedies which act more slowly, but more permanently, in relieving pain and the disposition to pain, we shall find that by far the greater number of these are such as any one must acknowledge to owe their efficacy to their power of promoting nutrition. Cod-liver oil and iron obviously act in this way; the latter by improving the quality of the blood, the former not only

9

in this way, but also, in all probability, by specially nourishing the nervous tissue. Quinine seems to have a double influence, at once improving the quality of the blood, and strengthening the systemic circulation. We are entitled also to suspect strongly that iodine and the iodide of potassium act by directly improving the nutrition of the blood; for any one who has watched the gradual abolition of the syphilitic cachexy under the influence of these remedies, must reason strangely if he do not arrive at the conclusion that these remedies effect a cure precisely at that moment, and at no other, when they have built up again the disorganized structure of the vital fluid. With regard to zinc, the testimony is very conflicting: and if we eliminate those cases in which valerianic acid has been used in combination with it, we shall find much difficulty in assuring ourselves that it is a true remedy for pain of any kind. There remain a host of remedies, the prolonged use of which has often proved useful, such as naphtha, æther, and alcohol, in repeated small doses, numerous volatile oils, and numerous vegetable substances containing such oils:—concerning all these remedies, it may be said that they prove useful in proportion as nutrition improves under their continued use, and no otherwise. The arguments for this view will be partially worked out in the special remarks on Anæsthetics, in the latter part of the present work.

The effect of the persevering use of cod-liver oil in removing neuralgic pains, is typical of the action of this whole group of remedies. The case above referred to, in which temporary relief was obtained by the use of alcohol, illustrates this action of cod-liver oil very forcibly, and may be taken as an example of what I have now witnessed several times. The patient was a man aged sixty, who applied for relief at Westminster Hospital, suffering from severe tic, which affected the branches of the second and third division of the fifth nerve on the right side. He had been treated with almost every imaginable sedative, without deriving more than the most trivial relief, and the disease was of eighteen months' standing. (January, 1862.) The face was haggard and worn, and there were frequent twitchings of the facial muscles on the affected side: the pulse was slow and rather weak, the appetite very poor, and the patient complained of great sleeplessness, which had latterly increased, owing to the increasing severity of the pain. The tongue presented a curious appearance, being strongly furred in its right lateral half, and comparatively clean as to the rest of its surface. The patient suffered much from cold hands and feet. Cod-liver oil was prescribed in increasing doses, till 3iss was taken daily, and this treatment was steadily persisted in for six months. At the end of about three weeks a great improvement was manifested, both as to general health, and also as to the pain; the patient also slept better: from this time forward, although there were many relapses, his tendency to improvement was progressive, and he was finally discharged cured, having been free from any attacks of the pain for five weeks.

II. The removal of muscular tremor, spasm, and convulsion, is one of the most striking effects which is produced by stimuli. According to the stock descriptions in treatises on therapeutics, they produce this effect by acting in this instance as "antispasmodics;" which is as much as to say that they relieve the symptoms by virtue of their ability to relieve them. In fact, upon the theory that all muscular movement is a result of stimulation or irritation, it was altogether inexplicable that convulsive movements should be arrested by the action of stimuli: so a sort of tacit agreement seems to have been formed to call the agents which were found to have this effect by a name which concealed the fact that they were stimulants.

Among stimulants which have the power to arrest convulsive movements, the first rank must be given to chloroform, administered in small doses by inhalation. It is commonly supposed that chloroform arrests convulsions by inducing a narcotic state, but this idea arises from an imperfect acquaintance with the order of phenomena in the induction of anæsthesia; for, in fact, true narcosis from chloroform is a state highly favorable to the production of convulsive movements. It is the state of stimulation produced when only a small dose has been breathed which puts an end to convulsive muscular movements; to go beyond this point would be to risk their recurrence, as will be shown hereafter.

A good illustration of the action of small doses of chloroform in cutting short convulsive attacks, fell under my notice some years ago, in King's College Hospital. A woman of middle age, who was the subject of epileptic fits, which recurred every day, and sometimes more than once a day, was treated by the administration of chloroform. At any recurrence of the convulsion, fifty minims were placed in a Snow's inhaler, and the weak atmosphere of chloroform vapor thus formed was breathed by the patient; a very few inspirations were always sufficient to subdue the convulsions, and to restore the patient to complete consciousness. I repeatedly administered the same dose myself, and always with the same effect: and extensive experience in the production of anæsthesia enables me to state confidently that no true narcosis *could* have been produced by so small an amount of chloroform, as indeed was sufficiently demonstrated by the collected state of the patient's intellect as soon as the convulsions ceased. Since that time I have very frequently adopted the same course with epileptic patients, and always with the same result: chloroform being carefully administered in the way above described, the convulsion always ceases. The influence of this treatment upon the ultimate progress of the malady appears, however, to be very small; though on this point more evidence is required. The convulsions of teething may also be arrested in the same way; and on the whole, I am inclined to expect better ultimate results from this application of the stimulant powers of chloroform, though it is necessary to speak with great reserve. With regard to the use of chloroform in tetanus, there is the same uncertainty as to *lasting* benefit; but there can be no doubt whatever that the muscular spasms may be resolved in most cases; and I am satisfied that it is the administration of a *small dose*, such as is quite insufficient to produce narcosis, which is best suited to this purpose. Some experiments will hereafter be detailed, on the action of chloroform in relieving the artificial tonic spasm of strychnia-poisoning in animals.

The action of small doses of chloroform in arresting convulsive movements, affords a good example of the operation of minute quantities of a substance, which in larger doses is an undoubted narcotic, or paralyzer. The operation of an undoubted stimulant, such as carbonate of ammonia, in a five-grain dose, produces smaller, but precisely similar, effects. We cannot administer remedies by the stomach during the progress of an epileptic convulsion; but in cases where the patient is in time to swallow a dose on the occurrence of threatening symptoms of a fit, the catastrophe is not unfrequently warded off. Dr. Reynolds* notices the fact that the epileptic fit may sometimes be retarded by the use of ammonia: and I am able to refer to a number of instances in which this has proved to be the case. In one patient, a man aged forty-two, who had been the subject of epilepsy for about fourteen years, the fits were warded off many times successively; so that at one period he absolutely passed a month entirely without fits; as he invariably took the dose of ammonia on the occurrence of premonitory symptoms. This man assured me, that the moment the sense of warmth induced by the ammonia spread over his system "so as to reach his finger-ends," all the unpleasant symptoms, including a spasmodic jerking of the right foot, vanished. After this long interval of repose, the fits began to return,† and the medicine was only occasionally effectual in preventing the occurrence of the fits. This patient afterward derived much benefit from

* On Epilepsy. Churchill.

† The patient's own vicious habits were partly responsible for this.

the persevering use of cod-liver oil; but it was a hopeless case as to prospects of cure.*

Intermediate between chloroform and ammonia, in the degree of its efficacy in arresting convulsive movements, is alcohol. I have notes of several cases of epilepsy, in which I was assured by the patients that they could sometimes ward off a fit completely by taking a tumblerful of hot brandy and water, or a glass of wine, as soon as any threatenings were perceived: and in two very severe cases under my care, alcohol has appeared to be absolutely the only remedy which was capable of mitigating the violence and duration of the paroxysms (chloroform could not be tried): in one case ammonia had completely failed, in the other it produced only trivial effects. In one of these cases I had the opportunity of personally observing the arrest of a fit, which was imminent, by a dose of brandy. I was conversing with the patient in his own house, when a curious expression of horror and bewilderment passed over his face, he began to stammer in his speech, and the head was jerked several times spasmodically toward the right shoulder. A hot glass of brandy and water was immediately administered, and the sufferer sat down quietly, the convulsive jerks ceased in a minute or two, and although the patient continued trembling and shaking for some time, there was no return of the threatening symptoms. The convulsions of teething in children form another class of diseases for which there is very considerable reason to believe that alcohol is one of the best remedies possible. Modern clinical observation has done much to discredit the views which were formerly held, not merely as to the sthenic character of the irritation which is present in these cases, but also as to the possibility of relieving it, either by general depletory measures on the one hand, or by incision of the gums on the other.

---

* While this goes through the press, a patient has come under my care, who has many times succeeded in averting a threatened fit by the inhalation of the vapor of liq. ammoniæ.

On the latter point some observations of Mr. Clendon* are worthy of the most serious consideration; for that gentleman argues forcibly the improbability that the *soft tissues*, by their resistance to the emergence of the tooth, are the cause of the nervous irritation; and gives good reasons for the belief that the inadequate space for the teeth in the jaw itself, and the pressure on the dental branches of the nervus trigeminus which results from this crowding, is answerable for the mischief. If this be the case, the evil is unremovable except by time, and meanwhile the object would seem to be to prevent the continuance of the convulsions, which might fatally exhaust the little patient's strength: and acting upon this idea, I have latterly given a trial to alcohol, in the treatment of these complaints. My experience is far too limited to enable me to speak with confidence; but in the three cases in which the plan has been tried by me, it has appeared to produce the best results; and at least I have little doubt that the convulsions may often be temporarily arrested by a dose of alcohol which is insufficient to produce any toxic effects. It need hardly be said that, supposing alcohol to be a suitable remedy at all, its rapidity of action is a strong recommendation in cases of this kind: and the same remark applies to chloroform.

A very curious variety, as it seems, of choreic convulsion, which is characterized by involuntary movements of a rotatory or semi-rotatory character, has been treated by Dr. Radcliffe, in two remarkable cases, by repeated doses of alcohol, with very complete success, the movements being arrested when a full stimulant effect had been produced.

It is needless to dwell upon the well-known effects of alcohol in relieving many kinds of muscular spasm: the most familiar instances, perhaps, are its beneficial influence upon colic, and upon spasmodic asthma; in both of which affections it often acts favorably: it is proper, however, to remark, that the use

* On the Causes of the Evils of Infant Dentition. By J. E. Clendon, M.R.C.S. London: T. Richards.

of intoxicating doses is entirely unnecessary, and, I believe, indirectly very injurious, by the after-depression which it causes. I would venture also to express the opinion, that the very sensible relief which is often experienced in spasmodic asthma from the inhalation of the smoke of tobacco, and of stramonium, may be procured without carrying their action as far so to produce a truly narcotic or paralyzing effect. This statement is made on the strength of observations made on asthmatic patients under my own care, as well as from what I can gather by conversation with medical men much accustomed to the treatment of this disease: and I cannot but think that the fact that asthmatic fits so constantly occur either during the feebleness of the nervous centers, which accompanies the state of sleep, or else during some artificially induced condition of depression, favors this idea.

With regard to that minor degree of involuntary muscular action which we are accustomed to speak of as *tremor*, it is scarcely necessary to say anything as to its amenability to the influence of alcoholic stimuli in many instances. Perhaps the most remarkable instance of this is the fact, that the very tremors which have been caused by narcotic doses of alcohol, are often relievable by means of small doses of the very same agent.

If now we leave the subject of the more rapidly acting anticonvulsive remedies, and proceed to consider the measures which are calculated gradually to eradicate the convulsive *tendency*, it is impossible not to see, at a glance, that these remedies are all of them probably—the great majority certainly—such as directly tend to improve nutrition. One by one the various sedative remedies appear to lose the confidence of the profession, that is to say, as far as regards their employment in sedative or depressing doses. Of the tonic medicines now adopted into more general use, by far the most effective is *cod-liver oil:* the experience of Dr. Radcliffe on this point is fully borne out by my own, as recorded in a paper read before the Western Medical and Surgical Society of London.* In the decidedly anæmic cases, *steel* sometimes acts excellently, when the disease is within reach of any treatment at all. Quinine has proved equally effective in a smaller number of cases,† more particularly in such as are characterized by some local numbness, either persistent or recurring in the form of an *aura* preceding the fits ; and the bromides of potassium and ammonium have been proved to possess an extraordinary influence in reducing the number of fits, even in very bad cases. All these are probably to be looked upon as agents which tend to restore nutrition to a healthy state.

III. Reduction of unduly frequent circulation. The most familiar instances which can be adduced of this result of stimulation, **is the** extraordinary influence exerted on the frequency of the pulse in febrile and inflammatory affections, by the administration of wine or spirit. So well known is this effect, that a certain degree of frequency of pulse, varying somewhat according to the observed type of the disease, is very commonly taken as the best indication of the necessity of administering stimulants: and granting, even, with Dr. Stokes, the importance of testing the strength of the heart's action by the audibility of its sounds through the stethoscope, nevertheless, on the whole, the mere frequency of action is the safest guide; for it will be found, **I believe**, in the large majority of cases to follow *pari passu* the gravity of the other symptoms, and it is certain that mere apparent strength of the heart sounds can by no means be always depended upon as annulling the indications of a rapid circulation. Here again it is by no means desirable to drench the patient with a large dose; the object should be to administer small quantities at short **intervals. For to** narcotize a fever patient, is a most serious and dangerous step; **and** the well-meant zeal of those who have desired to procure sleep, has often induced coma, from which the patient has only recovered to collapse and quickly sink.

* *Vide* Med. **Times** and Gazette, 1862.

† As an example of this, see a case published by the author in **the Medical Times and Gazette, 1861.**

Upon the reduction (IV.) of excessive secretion, and the relief (V.) of general debility, or of special fatigue of the brain, the muscles, or the digestive organs, as examples of stimulant action, I need not dwell here, for they are familiar even to triviality : but I would venture, here again, to insist that **effects** of this sort, when produced by a so-called narcotic, are **often** falsely ascribed to its narcotic action, when they are really owing to the fact that so small a dose has been employed as could only act as a stimulant. With respect to *excessive secretion*, I must be understood to mean merely **excess** in *quantity* of matter secreted, with the distinct understanding that I do not regard such a phenomenon as a **proof of exalted natural function, but as an instance** of morbid waste : upon this point I agree with the remarks of Dr. Chambers* and Dr. Inman.†

VI. The removal of delirium, and of maniacal excitement, and the production of healthy sleep where it has been morbidly absent, is an effect of stimuli which, according to the older ideas of stimulus, would surely have been unreasonable and impossible : and yet it is really one of the most characteristic of effects which these agents are capable of producing. So far as regards mania, and the majority of cases of delirium, this fact has at length been reluctantly acknowledged; but the occurrence of healthy sleep is still tacitly assumed to imply, of necessity, a subsidence of the bodily vigor, and accordingly we still hear of the necessity of "calming excitement," "reducing nervous action," &c., before sleep can take place, in cases of febrile disturbance. Upon this very important matter it will be necessary to dwell with some particularity.

That the brain does become depressed, to a certain extent, during sleep, in so far as this may be judged from its being in a state of *anæmia*, or bloodlessness, as compared with its condition in **waking moments, is rendered** probable by the

---

* Renewal of Life, p. 37.

† Foundation for a New Theory, &c.

recent researches of Mr. Durham.* But it is certain that there is a more extreme degree of anæmia of the brain which is absolutely incompatible with sleep of a natural and healthy kind. It was fully established by Dr. Todd, in his admirable Lectures on Delirium and Coma,† that the anæmic state is highly favorable both to the production of delirium and also of *coma :* and the history of fevers, of alcoholic delirium, and of the delirium and coma of uræmic poisoning, abundantly prove that there is a natural opposition between the tendency to these two formidable affections and the inclination to natural sleep. For the most part, that which will produce natural sleep will tend to check delirium and remove coma, and *vice versâ.*

It appears desirable to take as our starting-point the physiology of natural sleep, in order to explain the kind of stimulant action with which we are at present concerned. Natural sleep is that repose of the brain which, in a healthy state of the organism, and the absence of artificial hinderance, follows the performance of a certain amount of bodily labor, which has exhausted the nervous system to a certain extent. We may suppose, if we please (with Mr. Durham), that the results of the chemical changes in the nervous matter which necessarily accompany the action of the brain—the *débris,* so to speak—have impeded the continuance of the action by which they were produced, and that a state of comparative anæmia follows, during which the brain is allowed time to repair itself, as we know that it certainly does. Provided that this process of repair can go on properly, sleep is interrupted; not that it is a uniform state, on the contrary, as Sir Henry Holland justly observes,‡ it is a "series of fluctuating conditions, of which no two moments, perhaps, are strictly alike;" the variations extending "from complete wakefulness to the most perfect sleep of which we have cog-

---

* On the Physiology of Sleep. Guy's Hospital Reports, 1860.

† Lumleian Lectures; College of Physicians, 1850. *Vide* Med. Times, 1850.

‡ Chapters on Mental Physiology. 2d Edit. p. 6.

nizance from outward or from inward signs." It is a state which, as it seems to me, we may best express by supposing the mind to be in perfect vigor, but united to a corporeal instrument, whose efficiency is constantly fluctuating, in correspondence with the interstitial changes by which it is slowly being repaired, but with a steady progress on the whole toward recovery.

In delirium, on the other hand, whether produced by disease or by narcotic poisoning (hereafter to be described), in insanity, and in coma, an impaired condition of the brain is present which is *not* in the course of spontaneous rectification: the very gravity of the import of these phenomena lies in this fact. In order that the organism may pass from one of these conditions to that of natural sleep, it requires to receive some aid either from common food or from some stimulant which has the power to act like food. A typical example of this action is to be found in the treatment for delirium tremens, which in one form or another has long been adopted. Formerly the custom was to rely upon alcohol and opium: small stimulant doses of these agents not unfrequently cured the delirium and muscular tremors: if these failed, however, larger doses of the same remedies were tried: but rarely, I believe, with beneficial result. The narcotism thus induced was unnecessary, and not unfrequently fatal; its depressing influence being excited, not gradually and by slow degrees, but suddenly and powerfully, depressing the heart's power in so rapid a manner as to cause collapse and death. At present a more rational plan is substituted for this: the stimulant action of *food* is brought to bear upon the fatigued and anæmic brain, and the result, as I have myself witnessed, is the cessation of the delirium and the production of healthy sleep. Food is *the* stimulus, *par excellence,* for the brain which frequent narcotism has reduced to the state in which delirium occurs—as I need hardly say it is in the treatment of acute mania and for the wild violence of patients who are suffering from general paralysis of the insane. In all these cases the action of food may be supplemented or partially replaced by stimulant doses of alcohol, ammonia, &c., but *true narcotics are injurious.*

It is asserted by some, however, that delirium varies greatly in character, and that it presents, in different cases, features which are diametrically opposed, and requires an equally varying treatment, in accordance with these distinctions. There is a great improbability, I think, at the very first sight, in this view: nevertheless, it has the support of such high authority that it cannot be lightly treated. It is perhaps hardly necessary to notice what was said by authors who belong to an epoch when medical opinion tended to ascribe every kind of delirium to the effects of inflammatory disease: but the maintenance by Dr. Murchison, in his most excellent work on "Fevers," of the existence of a *delirium ferox,* which is opposed in its characteristics to the ordinary type of the affection, was to me a considerable surprise. This delirium Dr. Murchison describes as characterized by flushing of the features, loud talking, and violence of demeanor, a bold expression of countenance, &c. These features may belong to *delirium ferox,* but they have certainly also belonged to more than one severe case of delirium, in typhus, erysipelas, &c., which has fallen under my care, and which presented no exception to the ordinary rule which calls for the administration of support and stimulus. And here I would not only produce the evidence of my own personal experience; I appeal to those gentlemen who may have held clinical offices in King's College Hospital in Dr. Todd's time, whether such cases were not repeatedly observed, and as repeatedly treated with marked success by means of stimulants. For my own part, I may say with confidence, that the most typical case of delirium ferox which I ever saw was one in which slight delirium, occurring in the course of typhoid fever, had been accepted as a proof of encephalitis, and treated by leeching the temples, combined with starvation. The patient, who was young and muscular, was really rather an object of terror, so savage was the expression of countenance and so violent were the muscular efforts. Nevertheless, the yolks of half a dozen eggs, and half a dozen glasses of port wine, administered at intervals of an hour, vanquished the delirium completely. Nor did the results of this case surprise me, reflecting, as I did and had often done, on the incredible violence and muscular strength of many patients who are sinking into a state of general paralysis of the insane, and on their docility when treated with nutritious food.

The production of natural sleep in cases of fatigue carried so far as to render it impossible for the tired brain to sink spontaneously into repose, is one of the most valuable properties of stimuli. It is one of the commonest observations that severe and continuous labor, whether of body or of mind, will often produce a state of restlessness in which it is utterly impossible to sleep. Under these circumstances a small dose of opium will sometimes calm the agitation of the nervous system, and induce sleep; but more frequently the same object will be better attained by means of a cup of green tea, a glass of wine, or spirit and water, or a basin of good soup, the last being the best remedy of all. Another well-known remedy, which occasionally proves useful, is the immersion of the feet in hot water. And Mr. Durham has ingeniously suggested that this is because the brain, which should be anæmic in sleep, is congested, and the application of heat to the extremities acts as a derivative. It is certain, however, that the congestion theory will not explain the phenomenon of restlessness in many varieties of fatigue, nor the strictly analogous "jactitations" which occur after severe hæmorrhages and in the course of peculiarly depressing fevers, and it is probable that the influence of hot foot-baths is merely that of removing a present sense of discomfort, and also of producing a mild stimulant impression upon the nervous system generally. When intense fatigue is accompanied with great coldness of the extremities, this remedy is often very

useful in the way above suggested. The principle of cure in
all such cases is stimulation, whether there be greatly depressed circulation or not, and whether the feet are cold or
warm ; and the best stimulant of all, where it can be digested,
is meat soup, next to that of alcohol. Those who have not
tried the practice extensively would be surprised at the powerful effect which can often be obtained in the state of wakefulness accompanied by fidgeting movements, which is so apt
to occur in fevers, by the use of food in sufficient quantity.
The value of this remedy cannot be tested properly, in many
cases, unless the food be administered per rectum, as a sufficient quantity can seldom be tolerated by the stomach ; but
it is precisely these cases which are most strikingly bene-
fited—natural and refreshing sleep being frequently induced
within a very short time. With regard to the action of alcohol in such cases, it is to be noted that the quantity which
can be taken without producing more than a gentle stimulating effect, is often greatly increased, so that the patient will
take half-ounce doses of spirit, or glasses of port wine, in
rapid succession, without showing the faintest sign of alcoholic
narcotism, either at the time or afterward. Both food and
alcohol are preferable, as stimulants, to opium in these cases;
nevertheless, there are many fever-patients, and still more
erysipelas-patients, who are greatly quieted by minute doses
of opium quite insufficient to narcotize. In erysipelas, the
simultaneous use of brandy and small doses of morphia often
produces a series of short, sweet dozes, without the least
tendency to coma.

VII. The support of the organism in the absence of ordinary food, by stimulants, is one of the most remarkable phenomena which can be offered to the attention of the physiologist. Since the majority of substances which are capable of
acting in this way are also capable, in large doses and under
some circumstances, of acting as true narcotics, it has been
somewhat hastily assumed that the only manner in which
alcohol, opium, &c. can act as food, is by means of their capacity

for arresting vital changes and causing life to go on at a low degree of intensity.

Concerning this hypothesis, it is only necessary here to remark—first, that no proof exists that the measure of vitality ought to be taken from the rapidity of tissue-change, although tissue-change is necessary to life; secondly, that it has never been proved that arrest of tissue-change does result, at any rate to an important extent, as the result of the daily use of small quantities of the ordinary stimulant *ingesta;* thirdly, that observations are entirely lacking as to the rapidity of tissue changes in the extraordinary instances in which life has been preserved in the absence of ordinary food, by the action of a stimulant, for a considerable period; but that, fourthly, the recorded instances are numerous in which the latter occurrence has been noted, and in many of these vital energy would seem to have been maintained at a high point.

The effects now ascertained, beyond doubt, to be produced by the regular consumption of moderate doses of the coca (the great Peruvian narcotic-stimulant) by the natives of the country where it grows, are quite unreconcilable with any reasonable idea of narcotic action, and are clearly to be ascribed to the influence of a powerful stimulant. Von Tschudi,[*] Markham,[†] Pöppig,[‡] Weddell,[§] and many others, substantially concur in declaring that this wonderful drug, when taken in moderation by the South American Indians, or even by foreigners resident in the country, enables the taker to perform the most severe physical tasks upon an extremely small allowance of common food; and this without suffering any evil after-effects, such as invariably result from the use of any true narcotic which is not extremely volatile and eliminable. Von Tschudi, in particular, relates that an Indian, sixty-two years of age, worked for him (at excavation) for five days and nights consecutively, without any ordinary

---

[*] Von Tschudi. Travels in Peru, p. 451, &c.
[†] Markham. Travels in Peru and India, p. 232.
[‡] Pöppig. Reise in Peru, vol. ii, p. 248.
[§] Weddell. Voyage de la Nord de Bolivie, p. 516.

food at all, and with a very short allowance of sleep, and yet, at the end of that time, was fresh enough to undergo a long journey, simply because he was supported by the coca which he chewed from time to time. He declares that the moderate eaters of coca are long-lived men, and that they perform ex- tremely hard labor upon a very little food, as miners, soldiers, &c.; and he mentions the fact that the custom of coca-chew- ing is of immemorial antiquity in Peru. The true *narcotic* action of coca is quite a different thing, and the effects of it, as seen in the so-called *coqueros*, are quite as lamentable as anything we hear of the effects of excess in alcohol or opium. But the moderate use of the drug seems to have an influence upon nutrition almost undistinguishable from that of ordinary food as to its ultimate results.

Next, perhaps to coca, in its power of replacing ordinary food, we must reckon tobacco. The power of this substance to compensate, to a certain extent, the want of food, is very well known, but strangely enough it is generally assumed that this property of tobacco is dependent upon its power to disgust the appetite, by prostrating the nervous power of the stomach. A very little reflection should be sufficient to en- tirely discountenance such a view. There are very many substances capable of destroying appetite, by a depressing in- fluence upon the nervous system; such, for instance, as the salts of antimony, or the preparations of ipecacuanha; yet no one will pretend that the action of any such drugs would re- lieve the sense of faintness produced by fatigue, endured in the absence of food—an effect which tobacco undoubtedly produces in persons with whose system it agrees. The ex- perienced sportsman, accustomed to tramp long hours over the heather in quest of game, would laugh at such an expla- nation of the effect of his favorite "cutty:" he knows very well that it is by no mere disgusting of his appetite that he comforts himself for the indefinite removal of the prospects of dinner. By the time he had succeeded in depressing his stomach to the level of indifference to food, he may be sure

10

he would have rendered himself incapable for continued strenuous exertion, were tobacco effective only in this way. That
tobacco is not an exact equivalent for roast beef, nobody
knows better than the smoker; at the same time it would be
impossible to persuade any one, who had practical experience
of the use of it, to believe that its only effect is to depress
nervous power. The fact is, that all such statements are
made on the authority of persons either practically ignorant
of the effects of smoking, or else naturally incapable, as some
are, of deriving benefit from it. There are a few people whom
no amount of care and skill exercised in the taking of tobacco,
nor any moderation in the dose used, can save from unmistakable poisoning, whenever they indulge in it. These cases
are rare: and they ought to be carefully separated from the
evil results which are produced by mere *unskillfulness* in
smoking, such as causes the troubles of beginners in the art.

Next to tobacco in efficacy as a supplementary food, and
far surpassing it in its effectiveness under certain circumstances, is Alcohol. It would be opening far too wide a field
to attempt to discuss this subject here; and I may briefly
refer the reader to the researches recorded in the latter part
of this work, for the proof of what I must be content to assert
somewhat dogmatically, at present: that alcohol taken alone,
or with the addition only of small quantities of water, will
prolong life greatly beyond the period of which it must cease
if no nourishment, or water only, had been given: that in
acute diseases, it has repeatedly supported not only life, but
even the bulk of the body during many days of abstinence
from common foods: and that, in a few instances, persons
have supported themselves almost solely on alcohol and inconsiderable quantities of water *for years.* If these things
can be proved, as I shall hereafter show they can be, there is
no need, of course, to argue further about the alimentary
character of alcohol. We may be at a loss to explain the
chemistry of its action on the body, but we may very safely
say that it acts as a food.

I might strongly urge the claims of tea and coffee to be credited with the food-action upon the organism; but as it would be difficult to say how much of this belongs to their narcotic stimulant elements, and how much to the salts, and the tannic and caffeic acids which they also contain, it is, perhaps, better not to adduce them as examples of the food-action, which I am endeavoring to illustrate. I shall, therefore, conclude the series by a few remarks on the effects of this kind which are occasionally produced by opium.

Opium is one of those physiological agents which, by the mystery of their effects, excite much premature speculation as to their properties; and it is to this that we must ascribe the extremely incorrect descriptions of its action upon the organism which have appeared even in scientific works of good repute. One great source of confusion in the popular ideas of the action of this drug has been the disproportionate attention not unnaturally given to its soporific effects, which constitute, however, but a part, and that the least remarkable, of its action upon the system. In the countries where opium is indigenous, it is an article in daily use with the great majority of the population, by whom it is employed for a very different purpose than that of procuring sleep; in fact, as a powerful and rapidly acting stimulant: and in those localities far larger quantities can be taken without producing any other effect than this, than in the countries of Europe, where the poppy is only a transplanted growth. Taken in still larger quantities, even by the natives of Syria and the East, it proves as decidedly and poisonously narcotic as would much smaller doses taken by an Englishman; and this kind of effect is, doubtless, often seen as a consequence of the *abuse* of opium by Orientals. But its *use* is an important and genuine one: it acts as a powerful food-stimulant, enabling the taker to undergo severe and continuous physical exertion without the assistance of ordinary food, or on short rations of the latter—a fact to which numerous Eastern travelers testify. Dr. Barnes* relates a striking instance of its power to recruit the exhausted frame: "On one occasion I made a very fatiguing night-march with a Cutchie horseman. In the morning, after having traveled thirty miles, I was obliged to assent to his proposal of halting for a few minutes, which he employed in sharing a quantity of about two drachms of opium between himself and his jaded horse. The effect of the dose was soon evident in both, for the horse finished a journey of forty miles with great apparent facility, and the rider absolutely became more active and intelligent." Dr. Barnes declares that moderate opium-eating does not appear to shorten life or decrease vigor, an opinion in which he is supported by numerous competent authorities on the customs of the East— among others by Dr. Eatwell, who states that the health of the workmen in the opium factories is quite up to the average standard, and that the effect of the habitual use of the drug on the mass of the people (in China) is *not* visibly injurious. To a certain extent, and in certain circumstances, the same remarks would appear to apply to natives of this country, although the doses taken are, as a rule, much smaller than in the East. De Quincey mentions the fact that many poor, overworked folk, in towns like Manchester, consume regularly a moderate amount of opium; not using it as the means of a luxurious debauch, but simply to remove the traces of fatigue and depression : and the experience of physicians who know the poor of London would testify to the considerable prevalence of this custom among that class. It has frequently happened to me to find out, from the chance of a patient being brought under my notice in the wards of a hospital, that such patient was a regular consumer, perhaps, of a drachm of laudanum, or from that to two or three drachms *per diem*, the same dose having been used for years, without any variation. And I am assured that the practice is very extensively carried out in many parts of the country, just in this way, by persons who would never think of narcotizing themselves,

* A Visit to Scinde, &c., p. 230.

any more than they would of getting drunk; but who simply
desire a relief from the pains of fatigue endured by an ill-fed,
ill-housed body, and a harassed mind.   These instances appear to me inexplicable, except upon the supposition that
they depend on a kind of food-stimulant effect, similar to that
which certainly is experienced by the majority of Orientals
in taking opium; and they must be carefully separated from
that kind of narcotic delirium which is sometimes sought for
by the literary dilettante, and of which so vivid an account
has been left us in the "Confessions of an Opium-Eater."
The former effect is similar to that which I have myself experienced from a small dose of opium, as above related, when
suffering from depressing influenza; the latter is as truly a
poisoning as is alcoholic intoxication.   It must be added,
moreover, that the subjects are rare, in this country, in whom
there is any noticeable intermediate state between the stimulant and the narcotic action of opium; usually there is a
certain very small dose beyond which it is impossible to pass
without experiencing stupor, and the more prominent symptoms of narcosis.

VIII.   Local increase of nutrition.   This is a subject on
which we are as yet very much in the dark; nevertheless, it
is certain that the stimulant action of small doses of opium,
or of alcohol, often puts an end to a process of ulceration, in
which tissue is rapidly melting away, and enables the part to
heal.   It is customary to ascribe the extremely beneficial
influence of opium in sloughing phagedæna to its narcotic influence, but this is certainly incorrect.   When the opium
acts efficiently, not a trace of narcosis can be perceived: on
the contrary, the vital powers are distinctly raised; and the
cessation of the pain ought to be considered as a part of this
general improvement of power, and not as an indication that
nervous action has been "quieted."   And the same remark
may be applied to the beneficial action of alcohol in such
cases; for it is certain that there is no advantage in producing
even the earliest symptoms of intoxication in patients whom

we desire to rescue from the dangers of rapid destruction of tissue—on the contrary, such a practice is directly injurious.

There is another example* of what I believe to be a true increase of local nutrition as a result of stimulation, which is very generally overlooked by practitioners, or unknown to them : viz., the action of very small doses of strychnia. I shall have much to say hereafter as to the effects of *poisonous* doses of this drug, which I shall hope to show very ill deserves the name of stimulant, when given so as to produce tetanic convulsion. The effects I speak of are produced by doses altogether too small to induce tetanic convulsion, and were brought to my own notice, in the first instance, in a curious manner. A man, aged fifty-nine, applied for relief from an intense feeling of cold, attended with visible sluggishness of the capillary circulation, especially in the hands and feet, the remnants of an attack of hemiplegic paralysis, from which he had suffered twelve months previously, and from an extraordinary enlargement of the abdomen, which he had been led to believe was dropsical. The abdominal distention turned out, on examination, to be mere tympanitis, from partial paralysis of the muscular coat of the bowels; and small doses of strychnia were ordered ($\frac{1}{16}$-grain three times a day). On the occasion of the man's next visit, I thought at first that he was drunk, as he had the uncertain gait, meaningless smile, and flushed perspiring cheeks, characteristic of intoxication. To my surprise, however, I found that this effect had been produced by a dose of the strychnia taken half an hour previously, and he had come to me to complain of the medicine, because it "made him drunk:" this I ascertained, subsequently, by personal observation, was really the fact. The dose was decreased to the $\frac{1}{32}$ of a grain, and the disagreeable effects upon consciousness and co-ordination of movements were no longer observed; but a very remarkable increase of temperature of the surface was still produced by

* I have reason to believe, however, that this use of minute doses of strychnia is well known in India.

each dose, and by degrees became constant: the capillary circulation was restored to its natural activity, and a very considerable improvement took place in the nutrition of the muscles of the hands, forearms, calves, and soles of the feet. The abdominal distention, and the constipation and extremely troublesome tenesmus that accompanied it, vanished in about ten days from the commencement of the strychnia treatment; but the remaining symptoms were not completely cured till the medicine had been taken for nearly three weeks. At the end of that time he felt quite well, and was discharged; he begged that he might have the prescription, as the medicine, he said, was like a dram to him, if he felt cold and languid, and "there was no headache after it," at least when he took no larger quantity than the $\frac{1}{32}$ of a grain.

Since this case was under my care, I have closely studied the effects of minute doses of strychnia, and I am convinced that there is something which we do not understand, and which is completely distinct from the familiar effects of this drug, in its action when given in very small quantities. I am by no means sure that the effects do not vary, perhaps considerably, according to the taker's constitution; but this, at any rate, I believe to be a nearly universal rule, that doses which fall short of producing any poisonous effect increase the activity of the systemic circulation, diffuse a comfortable feeling of permanent warmth over the body, and favor the progress of local nutrition when this is deficient. The subject is too important and novel, however, to be disposed of by a mere analysis of cases and experiments, and I hope soon to publish the whole investigation at length. Such an effect as the pseudo-intoxication, which in the above-related case occurred when the dose was somewhat too high, was, I believe, an early symptom, and one which is probably very rarely seen, of the *poisonous* action of strychnia.

Such are the principal actions of *ordinary medicinal* stimulants, as they are to be seen in actual practice. Before we draw any conclusions, or even form any opinions from the facts thus briefly **presented,** let us reflect for a moment upon the action of a non-medicinal substance, which no one will deny is a stimulant of the first rank—namely, Oxygen.

If we were to **condense into** the shortest space the various therapeutic actions of oxygen, we could hardly express its efficiency better than by saying that it prevents or relieves pain, averts the disposition to muscular convulsion, tremor, and spasm, reduces excessive secretion, calms an unduly **frequent** circulation, removes general debility and special **fatigue of particular organs, quiets** the disturbed brain, compensates **in** great measure the absence of ordinary food, promotes local nutrition; in short, that it produces, with **tenfold** greater efficiency, all the results which we seek for by the **use of** medicinal stimuli such as those to which I have already **alluded.**

It needs but a glance **at** the vital condition of **different** populations in any country to arrive at a tolerably correct **idea of** the virtues of oxygen as a promoter of health and a curer of disease. If we compare the physical condition of **the** inhabitants of a London alley, an agricultural village, and a breezy sea-side hamlet, we shall recognize the truth of the description which assigns to it the same therapeutic action as is exercised by drugs, to which the name of stimulant seems **more naturally applicable** than to such a familiar agent as one which we are constantly breathing in the common air. A child that has been bred in a London cellar may be taken to possess a constitution which is a type of all the evil tendencies which our stimulants are intended to obviate. A Cornish fisherboy, such as Hook loves to paint, is the very model of **an** organization in which such morbid tendencies would be impossible. And even more remarkable, as showing the **power of oxygen** to avert the evils in question, is the immunity **from** them displayed, on the whole, even by so desperately **poor and** underfed a population as that of the southern agricultural counties of England, in consequence simply of the **out-door lives they lead.**

Now, if anything deserves the name of a *food*, assuredly oxygen **does, for** it is the most necessary element in every process of life. It is highly suggestive, **then, to find that** that very same quiet and perfect action of the vital functions, without undue waste, without hurry, without pain, and without *excessive* material growth, is precisely what we produce, when we produce any useful effect, by the administration of stimulants, though, as might be expected, our artificial means are weak and uncertain in their operation compared with the great natural stimulus of life.

It may be objected to any comparison instituted between the action of medicinal stimulants and of oxygen, with that of ordinary foods, that the effects produced by the two former differ from those produced by the latter in **this fundamental** point—that, in the one **case, the results produced are only temporary; in the latter,** they are permanent. **But this** argument appears to me to proceed on an erroneous assumption—the assumption, namely, that life is **to be measured by aggregation of** tissue, by activity of reproduction, or **by vio**-lence of dynamic manifestations. It is necessary to repeat that we have no right whatever to make such **a statement.** The nearest approach that we can make to a correct idea of vital perfection consists in the recognition **of the principle of** individuation—the formation **and maintenance of a certain balanced and** proportioned whole from **a multitude of parts, material and dynamic.** It is this balance—and not exuberance of **growth,** nor rapidity of tissue change, nor development **of heat,** or any other force, except so far as they minister to its preservation—which constitutes the ideal life of an **organism.** And, therefore, we should hesitate long before committing ourselves to the statement that there is a radical difference between foods and true stimulants, especially when we perceive that with regard to so many pathological and vital conditions their observed effects appear to coincide. It is certainly not for physiological chemistry, which is **still in** so rude and unformed a state, to insist upon any such stringent dogmatism.

Another grand argument against the propriety of comparing stimulants with true foods has always been, that *stimulus is invariably followed by reaction*. And physiologists have not been wanting who have ventured to apply to the phenomena of stimulation the positive law that "action and reaction are equal and positive." I need not dwell on the confusion of ideas which such an argument evinces; it is sufficient to point out the fact that, in giving the name "re-**action**" to the depressed condition which is supposed to follow the excitement of stimulation, we should apply an *active* epithet to what must be a mere cessation of action, on the theory that vital force is a thing *sui generis*, and can be *expended*. And besides, there is one more objection to be urged to the particular physiological dogma we are considering, which, I fear, must be as fatally decisive as was the officer's twentieth reason for neglecting to fire a salute in honor of Queen Elizabeth. Much might be said, and very forcibly, in **refutation of the** assertion that there is a logical *necessity* for stimulation to be followed by a recoil equal in extent to the first elevation; but we may save ourselves all this trouble by declaring simply that, in fact, *no such recoil occurs*.

The origin of the belief that stimulation is necessarily followed by a depressive recoil is obviously to be found in the old vitalistic ideas. It is our old acquaintance, the Archæus, whose exhaustion, after his violent efforts in resentment of the goadings which he has endured, is represented in modern phraseology by the term "depressive reaction." This idea, once being firmly established in the medical mind, the change from professed vitalism to dynamical explanations of physiology has not materially shaken its hold. · We speak of the exhaustion of muscular force or the exhaustion of nerve-force, not knowing really whether there be such *special* forces at all, **but** availing ourselves, only too readily, of the opportunity of giving a semblance of scientific accuracy to the expression of our prejudices.

It is not true that stimulation is of itself provocative of subsequent depression; but there are circumstances in which this might easily appear to be the case. For instance, when the superabundant mental energy of a man, whose physical frame is weak, induces him to make violent and continued physical efforts, he **is** apt to find, at the end of a short " spurt " of exertion, that his energy is exhausted. But here the exhaustion is no *recoil* from a state of stimulation, it simply represents **the** fact that the elements of the muscular tissue are disorganized, and that the attempt to put them in motion fails, creating, **at the same** time, a mental feeling of weariness. And the case of drunkenness—that is, of alcoholic narcotism—affords another excellent example of the fallacy we are considering. The *narcotic* dose of alcohol, as will **hereafter be shown,** is alone responsible for the symptoms of depressive reaction. Had a merely stimulant dose been administered, no depression would have occurred, any more than depression results from such a gentle use of the muscular system as is implied in a healthy man taking a walk of three or four miles. What depression is there, as an after-consequence, of a glass or two of wine taken at dinner, or of **a glass of beer** taken at lunch, by a healthy man? What reaction from a teaspoonful of sal-volatile swallowed by a person **who** feels somewhat faint? What recoil from the stimulus of heat applied in a hot bath, or of oxygen administered, **by** Marshall Hall's process, to a half-drowned man? *Absolutely none whatever.* The visible immediate results of these measures do, indeed, after a time, disappear, not being exempt from the ordinary conditions of temporal things; so that, just as food requires from time to time to be renewed, so does the oxygen which has been artificially driven into the drowned **man's** thorax require to be renewed by his own respiratory **efforts,** when he has once recovered the power to make any: **and so** does the **glass of** wine which we took to-day to relieve our sense of fatigue require to be repeated to-morrow, when similar circumstances present themselves as on that occasion.

We often **hear** the effects of strong irritation of the skin, or the **mucous** surfaces, quoted as an example of the way in which **action and** reaction follow each **other.** The immediate effect of such treatment (it is said) is to **quicken the circula-tion and improve the** vital condition **of the part, but its** *ulti-mate* **result is a** complete stagnation of the vital activities in **the irritated tissues.** The real **explanation of the matter is, however, very different** to this. Mild stimulation of the skin **(as by friction, warm liniments, &c.) has no tendency to** pro-**duce subsequent** depression; nor has **mild** stimulation of **the mucous** membranes (as by the **mustard** we eat with **our roast beef).** But the application of **an irritant strong enough to** produce a morbid depression at all, produces it *from the first.* **Thus the cantharidine of a blister has no sooner become ab-**sorbed through the epidermis than it *at once* **deprives a cer-**tain area of tissue of **its vitality to a considerable extent, as** is explained by the researches **of Mr. Lister, already referred to.** Here is no stimulation first and depressive recoil after-ward, but unmitigated depression **from the first.** But in the tissues more distant from the source **of damage, we find** the circulation proceeding more actively, and **some of the vital processes** apparently more active than in **health: the** general strength is, however, certainly much *reduced.*

The effect of tobacco-smoking, in moderation, on **the ma-jority of persons who are skilled** in the use of the pipe **is a marked instance of** stimulation, the pulse being slightly in-**creased in frequency,** and notably in force,* and **the** sense of **fatigue of body or mind being greatly relieved.** This stimu-**lation most assuredly is not succeeded by** depression. On **the contrary, the smoker feels lighter and more** cheerful, and the pulse maintains its firmness **in many** cases for an hour or **two, and even then yields to** no morbid depression. Where **depression is produced it** occurs *early,* **and is a sure sign that**

* *Vide* Dr. E. Smith's Paper, read before the British Association, 1862.

even the small dose is too much for the smoker's constitution, and that he had better not smoke at all. Of course this does not include cases where mere unskillfulness causes too large a dose of the constituents of tobacco-smoke to be taken into the system, as with nearly all beginners.

And finally, I may appeal to the experience of those who have taken opium, in a single small dose, when suffering from the severe depression of catarrh, whether any depressive reaction follows the delicious calm and the sweet natural sleep which is usually produced by this remedy. I believe that this is never the case unless the drug has been given in too large a dose, or the patient's constitution be one to which opium is obnoxious. If this unfortunately happens, then the true stimulant action may be induced (supposing absorption to go on slowly), while as yet only a small quantity entered the circulation. But on the full dose obtaining access to the blood, and through this to the nervous system, a state more approaching coma than natural sleep is produced, and on recovering from this the patient is made painfully aware (by feelings of languor, headache, &c.) that he has been narcotized. But this condition is not an after-result of stimulation.

In short, if it be sought to separate stimuli from true foods on the ground that the former produce a subsequent depressive reaction, such ground is untenable. A stimulus promotes or restores some natural action, and is no more liable to be followed by *morbid* depression than is the revivifying influence of food.

And if it be sought to distinguish foods by the peculiar characteristic of being *transformed in the body,* then I answer that this is the worst definition of food that can be given, since water, which is not transformed in the body at all, is nevertheless the most necessary element of nutrition, seeing that human life can only be maintained a day or two without it, but may subsist *for weeks* on water as its only pabulum besides the atmosphere and the tissues.

The one important difference which really does exist be-

tween stimulant drugs and the substances ordinarily called
foods is, that the former make no considerable positive excre-
ment to the bulk of the tissues. But, as we have said before,
life is not growth nor change merely, nor do we know at all
accurately what the conditions are under which it *may* be
maintained. We know that, in ordinary circumstances,
human life cannot be supported without food of a certain
quantity and quality; but we are by no means certainly informed whether this is always the case, as will be evident
from the facts already mentioned as to the effects of coca and
of opium, and still more from those to be related hereafter as
to the habits of some alcohol-drinkers. Reflecting on the
latter especially, as more within our immediate observation
in this country, it is difficult to avoid the conclusion that the
word "food" requires to have a more extended significance
accorded to it than is usually given, and that we must learn
to think it possible that circumstances may alter the conditions of life far more extensively than is allowed in ordinary
physiological treatises.

And I may be allowed to say that, supposing stimulants to
have a real affinity to foods in their action on the organism,
there is something in the fluctuating character of their action
which would seem to be appropriate, and such as might be
expected. It would be natural, on this theory, to suppose
that different morbid conditions might render the use of different stimuli effective, according as the one or the other was
present, just as various kinds of food become, each in its turn,
the most fitting for the support of life, in the circumstances
of ordinary health, according to varying age, sex, climate,
occupation, &c.

I venture to think that the considerations advanced in the
present chapter go far to justify the following proposals:

1. That the use of the word "stimulant" be restricted to
agents which, *by their direct action*, tend to *rectify some
deficient or too redundant natural action or tendency.*

2. That agents which produce excessive and morbid action

of any kind in the organism, be refused the name of stimulants, even though smaller doses of them may act in a truly
stimulant manner.

3. That the word "overstimulation" be entirely rejected
from use, as unphilosophical and a contradiction in terms.

I make no apology for having kept the argument clear of
any reference to the researches of physiological chemists; for
if the reader has followed the course of thought which has
now been developed, he will, I hope, perceive that the questions raised in the present and preceding chapters are, in
truth, preliminary, and require to be answered, as well as
may be, before we can make any useful application of
chemical research.

CHAPTER IV.

THE DEFINITION OF NARCOSIS.

In order clearly to explain the sense in which this word will be used in the present work, it is necessary, in the first place, to inquire a little into its ancient meaning, and also as to its convertibility with certain other terms, viz.: Nepenthism (or Anodynism), Sedation, and Hypnotism.

The original meaning of the word Narcosis seems clear. Hippocrates used the verb ναρκόω in an active sense—*to benumb or deaden,* and νάρκωσις for the *process of benumbing;*[*] sometimes also, apparently, he used the latter word in a passive sense,[†] interchangeably with νάρκη, to signify the state of numbness or paralysis. Erotian says that ἡ νάρκη is the suppression of feeling;[‡] and Aristotle uses this word to express the sensation of the hand or foot being asleep. The verb ναρκάω was used in a passive sense by Homer,[§] and by Plato,[||] &c.

Galen appears to have coined the word ναρκωτικός (whence our "narcotic"). He says[¶] that the agents which are thus to be denominated act by extinguishing or repelling the native heat and the animal spirit, without which feeling is

---

[*] Of the power of cold, he says, ὀδύνην ναρκοῖ. De Liquid. 42. (So also Alex. Aphrodis. Ναρκοῦντα τὴν δύναμιν αἰσθητικήν.)

[†] νάρκωσιν γνώμης, Epid. 6. 1. 13. νάρκωσιν κοιλίης, Epid 6. 3. 1. νάρκη μηρὸν, Epid, 6. 1. 5; elsewhere νάρκη ἐν τοῖς σκέλεσι.

[‡] See the first volume of the Geneva folio edition of Hippocrates' works; Erotian's Commentary. ἡ νάρκη θλίψις αἰσθήσεως ἐστίν.

[§] Iliad VIII. 328. Of Teucer, when struck by a stone, νάρκησε δὲ χεὶρ ἐπὶ καρπῷ.

[||] Meno, 80 B. In reference to the effect of Socrates' discourse upon him, Meno says, ἀληθῶς γὰρ ἔγωγε καὶ τὴν ψυχὴν καὶ τὸ στόμα ναρκῶ.

[¶] De Symptom. Causis. lib. ii, preface.

impossible. The chief examples which he gives* are the root of mandragora, the seeds of altercus, and the juice of the poppy : they are, however, dangerous remedies, and may be most safely employed in conjunction with aromatic stimulants, as in colic, for example.

That the word narcotic was used by the ancients to designate agents which possessed a paralyzing power, is certain. Sir W. Hamilton speaks very positively† to the effect that they understood this paralyzing process to extend only to the nerves of sensation; but it would, I think, be difficult to reconcile this idea with several expressions both of Hippocrates and of Galen, or with the passage above quoted from Homer. It seems also more especially contradicted by another part of the passage already cited from Galen.‡ It would appear more correct to say that in its original significance, narcosis meant a state of partial deadness and want of power both to feel and to move, but still short of what is produced by apoplexy, or by complete paralysis.

It is, I believe, a serious mistake to confound with narcosis the kind of effect which the ancients sought for from the administration of the various substances to which they gave the poetic name of Nepenthes. It is obvious that the idea conveyed in the poetic description of Homer, which was the origin of the medicinal use of the term, is that of a simple exhilarant,§ a remover of care and pain; and whenever we find mention of nepenthic remedies among the ancients, they are spoken of in this way, and no hint is given of their possessing a narcotic or paralyzing influence. The distinction made in early times between that sort of physiological action which merely quiets grief or physical pain, and the truly

---

* Hyoscyamus, "henbane."

† Notes on Reid, p. 869.

‡ De Sympt. Caus. loc. cit.

§ Odyss. IV, 220. Helen, in order to remove the sorrows of Menelaus and his friends,

αὐτίκ' ἄρ' εἰς οἶνον βάλε φάρμακον ἔνθεν ἔπινον
νηπενθὲς τ' ἄχολον τε, κακῶν ἐπίληθον ἁπάντων, &c.

11

poisonous or **deadening action** of narcotics, is well exemplified by Galen, whose **language is** very clear and explicit. In his chapter **"Quod anodynorum et narcoticorum** facultas," he **says*** that the remedies which should be called anodyne are even opposite in their qualities to narcotics: the former are of a hot and subtle nature, as is necessary in order that they may digest, rarefy, extenuate, and render equable, something which is in the parts vexed with pain, either of acridity, slowness, or thickness of the humors, or something hinders the free exit of the cold spirit from the part. But those substances which are given to procure sleep (hypnotics) are of a nature contrary to that of anodynes. For these all refrigerate the body, and so *narcotize* the sense, that if taken in a very large dose, they would cause death; and he speaks of some of the more formidable narcotics as causing madness or delirium, yet even these, in a convenient dose, and mixed with other things, may sometimes assist the cure of disease.

The Arabian writers follow generally the same theory as Galen held, and speak of true narcotics as agents opposed to those which enliven and quicken the vital spirits. Avicenna mentions† seven varieties of exhilarants, none of which appear, by his description, to have any narcotic qualities. And Palamedes, the annotator of Avicenna, distinguishes‡ between that kind of relief to pain which is given by such agents as Anethum (the oil of which is mentioned by Galen as a type of anodynes proper) and that other kind which is produced by agents which moisten and induce sleep and interfere with the natural processes, such as *the inebriants, and indeed all the narcotics.*

Of the various substances to which the poetic term nepenthes was applied, some were of the class which we should now call antispasmodic stimulants (*e.g.* mastic—the helenium odorum of Theophrastus, as some suppose, and elecampane),

---

* De Facultatibus Simp. Medicament. lib. v, cap. 9.

† Avicenna. De Viribus Cordis.

‡ Palamedes. Index in Avicenn. libros.

and some were undoubted narcotics, but used in such doses as from the description certainly did not produce narcotic effects, but only a mild stimulation. The most important of these were probably the Indian hemp, a plant very anciently used in Egypt, whither it had been brought from the East, and opium, also very anciently used in Egypt and still earlier in Oriental countries. Galen mentions* the seeds of hemp as in use in his day, as a sweetmeat for after-dinner use, and speaks of it as having a heating effect (which would place it in a different class from the narcotics). But we find authentic mention of hemp as used for its exhilarant qualities far earlier than this, by Herodotus, who says† that the Scythians inhaled its vapor. Diodorus Siculus speaks‡ of the women of Thebes, in Egypt, as having been acquainted from time immemorial with the use of a plant which had power to assuage grief, and conjectures that it was this very plant which Homer says had been given to Helen by an Egyptian woman, and from which she made the pharmakon nepenthes. If there really were this distinct origin for the Homeric legend, the plant in question may have been either the hemp or the poppy.

On the whole there seems good reason to consider that the idea implied in the terms Nepenthes, Anodyne, Exhilarant, like that of the sacred writer when he speaks of "wine that maketh glad the heart of man," is a just and important one; in fact that it is neither more nor less than the notion of *a highly diffusible stimulus.* That such an agency as this should be thought of as appropriate to the relief of pain, by Greeks at least, was natural enough, since Aristotle had held§ that pain was the reflex of bodily imperfection, and whatever removed imperfection removed pain, and Plato had spoken‖ of pain as caused by an interruption of the bodily harmony. All

* De Aliment. Facult. lib. i, cap. 34.
† Herod. lib. iv, cap. 75.
‡ Diodorus Siculus, lib. i, cap. 97, l. 7.
§ Nic. Ethic. lib. x, cap. 7.
‖ Timæus, 39. Republic, lib. ix, &c.

this has been in great measure lost sight of in later times, and nepenthic or anodyne action has been generally confounded with narcotism. It is time to restore the latter word to its true and original signification; for, in the present state of our knowledge, we shall hardly find a better definition of narcotics than that of *deadening agents, which diminish the activity of the nervous system.* Other actions may be, and probably are, produced by various members of this group, and especially it is probable that many of them cause important changes in the blood. But our knowledge of physiological chemistry is far too limited to enable us to trace these with distinctness; while, on the other hand, the definition above given is very useful as a hypothetic center from which to extend our inquiries as to the physiological action of individual narcotics, and to classify their varieties.

In the next place, an inquiry must be made as to the propriety of separating, as is now usually done, Narcotics and Sedatives. Notwithstanding the very high authority which has sanctioned this division, I cannot perceive the necessity, or even the advantage of it, while I am certain that it is most exasperating to the thoughtful student who has carefully studied the action of remedies at the bedside. The separation of narcotics from sedatives is sanctioned by Dr. Billing (followed by Dr. Pereira), by Dr. A. T. Thompson, Dr. Headland, Drs. Ballard and Garrod, and Dr. Neligan; and it may be said, without exaggeration, that we have in their remarks on this question no less than five distinct and irreconcilable accounts of the matter. It has been said that *sedatives* only affect the sensory nerves,—that they only affect the heart, leaving the brain and spinal cord untouched,—that their operation is depressing from the first, never stimulating at all,—that they prevent altogether the elimination of nerve-force: while narcotics act on the brain and spinal cord,—are stimulating at first and depressing afterward (on account of the recoil from the stimulant action, says Dr. Thompson, which Drs. Ballard and Garrod contradict),—do not alter the quantity of nerve-force generated, but only impede its communication (Billing),—act slowly as compared with true sedatives. With all this confusion and uncertainty as to the *way* in which sedatives and narcotics differ from each other, there is, at least, some agreement hitherto in placing certain well-known agents definitely in the one class or the other; for instance, tartar-emetic, digitalis, and tobacco* have been almost unanimously claimed as unquestionably belonging to the sedative class—their action being attended with no stimulant effects whatever, but being wholly depressive from the first, more especially to the heart. It is a curious reflection, that on the one point on which the authorities were agreed they were decidedly wrong. For with regard at least to tobacco and digitalis, it is now certain that those substances administered in certain doses and in certain circumstances act as powerful stimulants;† while of tartar-emetic, it is not unlikely that a part of its action is exerted on the central nervous system. Again, when chloroform was first applied to the purposes of operative surgery, it was claimed as a decided example of the sedative class; yet nothing is more certain than that it acts in small doses as a powerful stimulant, as is seen in its remarkable effects in forwarding the progress of parturition.

The one difference which would seem to stand out at all prominently as a distinctive mark, separating sedatives from narcotics, is the sudden and *shock-like* manner in which the former depress the system when given in a very considerable dose. Prussic acid and tobacco might be taken as typical examples of agents which are capable of occasionally producing this sort of effect; but neither of these substances is devoid of the power of stimulation when administered at a proper time and in a suitable dose. And on the other hand, it is certain that opium, which was selected by Brown as the

* But Dr. Headland calls tobacco an *inebriant.*

† This fact, though strangely lost sight of in recent times, was well known to older writers. See Gerard's "Herball," edited by Johnson, 1552.

most perfect example of a diffusible stimulant, has not unfrequently produced precisely that sudden depression which
should result only from the action of a sedative. It has
happened to me, as I doubt not it has to others, to see
patients suffering from delirium tremens, who had been in-
judiciously pressed with large doses of opium, because smaller
ones had failed to procure sleep, die from collapse almost as
suddenly as if they had been shot, or had swallowed a large
dose of prussic acid, most obviously from the effects of the
medicine.*

The most important of the phenomena of this kind of shock
is sudden depression of the force of the circulation, and it is
a result which has usually been considered characteristic of
the action of *sedatives;* but in truth its occurrence or nonoccurrence in any case depends very much upon the quantity
of the drug which is administered and the circumstances
under which the administration takes place. With regard
to chloroform this is particularly the case, as will be hereafter shown; but it is, perhaps, even more characteristically
displayed in the behavior of oxalic acid; for it is only when
this drug is introduced into the circulation in a diluted form,
and a considerable dose, that the paralysis of the heart noted
in Dr. Christison's experiments† is produced. When the
dilution is still considerable, and the dose smaller, the ordinary course of narcosis, including stupor, and not unfrequently 'convulsions, is passed through, and death at last
results from the cessation of the respiratory movements, or
else from the respiratory muscles becoming fixed in a violent
tetanic spasm, thus inducing suffocation.

The truth is that sudden depression of the heart's action
is an effect capable of being produced, under certain untoward
circumstances, by a variety of narcotics, and it may probably
be produced in either of two ways: 1st, By a depressing in-

* Dr. G. Johnson, who has kindly favored me with his views on this subject, insists
very strongly on this evil influence of large doses of opium in delirium tremens.
† Christison on Poisons, p. 219.

fluence on the brain of such an overwhelming character as to resemble the violence inflicted by a blow which completely crushes the head; and 2d, By the rapid production of a change in the organic nerves of the heart itself, through the direct action of poisoned blood upon them. But in whichever way it is caused, such a phenomenon as the very rapid sinking of the heart's action below the standard of normal frequency, is by no means to be looked on as in itself a medicinal or salutary but as a strictly poisonous effect. To a limited extent it may be admitted as a necessary sacrifice, as in the momentary collapse which necessarily precedes the desired action of certain emetics; but it is surely a most objectionable proceeding to constitute a class of remedies, under the name of sedatives, giving it to be understood that the central feature of their action, upon which their therapeutic value depends, is that of depression of the heart's force. The mere reduction of *abnormal* frequency is, as I have already stated at length, to be accomplished by *stimulants*. There is, as it seems to me, no room for such a special class of remedies as sedatives at all—the very word is objectionable, since its fundamental idea is derived from the notion of a forcible pacification of some tumultuous and disorderly demon within the organism. The phrase, therefore, will not be used in the following pages as indicating anything distinctive, but, if at all, only as a synonym for narcotics.

After what has been said in the section on Stimulation, the reader will understand that Hypnotism, or the production of sleep, does not form, in my opinion, a part of the rôle of narcotics as such. That it should have been considered so by the ancients was natural, because of its resemblance to *coma*, an undoubted result of true narcotism; and also from the fact that sleep is often rendered possible by small doses of the very same substances (*e.g.* opium), which in larger quantities produce coma. True sleep is, however, a part of the cycle of daily natural processes, and

can only be induced, if it be morbidly absent at a time when
the system has need of it, by remedies which reinduce a
normal state of the physical organism—by stimulants, in the
sense in which I understand that word. Hypnotic or soporific influences will, therefore, not be reckoned by me among
the phenomena of narcosis.

Narcosis proper.

Narcosis proper may be described as a physiological process
in which the nervous system is deprived, by the agency of a
poisoned blood-supply, of its vital characteristics, with greater
or less rapidity, and which directly tends to produce general
death of the organism by means of such deprivation. Its
varieties may be considered as dependent upon the order in
which the devitalizing power affects the various portions of
the nervous system, and the comparative violence with which
it injures them respectively.

As has been already remarked, Life is not mere nervous
action. But if the highest notion which we can conceive of
life is that which makes it consist in the individuation which
unites many forms of matter, and the various forces developed
among them, into one whole, of which we say that it is
"alive"—that it is "a living organism,"—then certainly we
may conclude that the nervous system is a prime instrument
in that sort of individuation which we call "human life." Its
structure and arrangement suggest to the most thoughtless
person the idea of connection: it seems created to be the
medium of interdependence for all other parts of the frame,
and more especially for that kind of interdependence which
demands an incessant instantaneous communication. The
messages carried by the blood from one organ to another
travel more slowly, nor do they concern matters of such
suddenly vital import—to speak for a moment in figurative
terms. To use plainer language, we may well believe, since
we know every part of the frame to be the seat of electric

phenomena, that the mere circumstances of its homogeneity, *through long spaces and many branches*, enables the nervous system to continue the most important and the slightest electrical disturbances, originating at any one of its terminals or on any part of its track, to an indefinite distance, and *with immediate completeness of effect.* And how much is implied in this possibility of the unlimited extension of electric change, is only beginning to be understood since we heard of the doctrine of the mutual convertibility of forces.

Narcosis may be, therefore, understood to be no less than the severance of the copula of life,—a severance partial or complete, according as it cuts through some mere solitary border-path, or the busy cross-roads of the ways of life; that is, according as it touches some outlying nerve only, or poisons the great centers through which they all communicate,—partial or complete, too, according as it partly obstructs, or wholly closes up, these paths of dynamic influence. It is, in fact, a more or less complete paralysis of the nervous system. We propose now to enumerate the principal groups of symptoms which may be produced by narcotic agents.

The general symptoms of narcosis may be divided into six classes, according to the particular functions of which they indicate the disturbance—viz. those of the Mind, of Sensibility, Muscular Movement, Secretion, Circulation, and Respiration.

I. The mental disturbances to which the action of narcotics may give rise, are as follows:—Loss of the reasoning faculty, of the moral sense, and of the power of voluntary recollection; prominence of the emotional and appetitive instincts; delirium, involuntary memory, and involuntary fancy; partial or total loss of consciousness.

II. The disturbances of sensibility include:—Delusive feelings of heat or cold, partial numbness, formication (creeping sensations), painful tingling, or actual continuous pain; indistinctness of the sense of touch, perversions of the other special senses; actual paralysis of common sensation, in most instances commencing with the posterior (in man the lower) extremities.

III. The **muscular affections** include:—Chronic convulsions; tremor and **shudderings**; **spasm**; tetanic convulsions; catalepsy; motor paralysis.

IV. The disturbances **of secretion are**:—Copious effusion of certain elements (especially **the watery), of one** or more **secretions, usually** accompanied **by the** non-elimination or **the alteration of the** other ingredients, **or** more or less complete **arrest of** a particular secretion, or of secretion generally.

V. The **alterations of** circulation are:—Undue frequency **of the heart's action, with abnormal** feebleness; undue frequency, **with undue force;** undue slowness, **with** abnormal **feebleness; irregular rhythm of the heart, with** abnormal **feebleness; changes in the pupil.**

VI. The **alterations of respiration include:**—Undue frequency; **undue slowness;** gasping, sighing, **laboring, or** spasmodic **respiration.**

The above **list includes all the nervous** phenomena of narcosis, **so far as we know them. In** the **case,** however, of particular narcotics, **this or the other** symptom **is** not produced; **and the respective** proportions in which they present themselves vary greatly in **the** case of different agents, and **also in** accordance **with the** amount of the dose, the rapidity **with which it enters the circulation,** the peculiarities of the **individual organism acted upon, and** the state of health at **the** time **of administration. In** all cases where narcotism is **carried to a fatal** termination, symptoms belonging to each **of the six** groups above referred **to** necessarily occur, since **this is** implied in the mere **fact of** death; but the order of **their occurrence is very different, according** to the agent used, and the **varying** conditions **of administration.**

It would **be** difficult to find **a better** example of the great variety **of** symptoms which the same **narcotic** may give rise **to, than by** sketching the various lines, **so to** speak, along **which** acute opium-poisoning may advance, to the extinction

of life.   In the human adult subject, the following is the
usual course of the symptoms:—so long as the blood has been
impregnated only with a small dose, no other effect than that
of a mild stimulant is experienced; after a time, however,
the patient experiences a sense of languor, and an inclination
to repose; if he yields to this, and lies down, he immediately
passes into a state which at first sight resembles sleep, but
which is really a different condition, namely, the commencement of a sort of coma.   It may be distinguished from sleep
by the fact that the patient may be roused, at first pretty
easily, by a shake, or even a question addressed to him in a
loud voice, but instantly relapses, without an effort to retain
his senses, into his previous torpor.   There is something
mysterious in the sudden way in which the poisoned person
appears to pass at once from a state of intelligent wakefulness
into this comatose condition; but from what I have observed
in a case of accidental poisoning with opium, the early symptoms of which came under my notice, it would appear probable
that consciousness has been considerably impaired before the
patient yields to the desire of lying down; but that a certain
automatic co-ordination of the movements, sufficiently perfect
to elude superficial observation, has continued.   By dint of
strongly rousing the patient's attention, we enable him, for a
moment, to concentrate his power of will upon the one object
of replying to a question, or of executing some movement
which we desire him to make.   And even when intelligence
seems lost, we still find that for some time strong physical
impressions made upon the surface of the body will compel a
mechanical performance of the accustomed movements of
locomotion, &c.   The progress of the symptoms, supposing
that recovery does not take place, is all in one direction, that
of ever-spreading paralysis.   The nerves of common sensation
lose their power of transmitting impressions; this phenomenon
is usually noticed first in the posterior (in man, the lower)
extremities, and travels slowly forward (or upward).   Supposing the affection of the brain not to proceed with such

rapidity as speedily to blot out consciousness, the paralysis of the sensory nerves is usually preceded by a stage in which there is *pain* of a dull aching character. By the time that sensory paralysis is developed to any considerable extent, there is a diminution in the power of the respiratory movements:* simultaneously with this, the pupil is observed to be strongly contracted and insensible to light. It is usually at this stage, if at all, that mental disturbance is evidenced by the presence of delirium: this delirium is apparently distinguished by the confused presentation of ideas or scenes formerly impressed upon the mind: the patient mutters, as in delirium tremens, and his words not unfrequently relate to events long past: but unlike the delirium of chronic alcoholism, this condition seems rarely to be accompanied with spectral delusions of a terrifying kind. In a further stage the coma becomes complete and profound, and the patient can no longer be roused, the respiration becomes more and more labored, the pupil becomes dilated, still remaining insensible to light, the countenance deadly pale; while a copious sweat breaks out, and the bronchial tubes become filled with mucous secretion, which increase the embarrassment of the breathing. **Involuntary** expulsion of urine and fæces often take place at this time. Finally, the respiratory power becomes progressively weaker; the breathing is now irregular, gasping, and labored, and at last ceases; the heart continuing to act for some little time, it may be for some minutes, later.

Such is the usual progress of acute poisoning with opium in the adult human subject under ordinary circumstances of general health. The variations which are presented by the action of this agent in other animal organisms, and in the human being, under certain circumstances, are highly interesting. If we examine the phenomena of opium-poisoning in infants, we are at once struck with the great rapidity with **which** profound coma is induced, and with the strong tendency to clonic *convulsive movements* which manifests itself toward

---

* The *frequency* of respiration under these circumstances is very variable.

the later stages. The great majority of recorded cases, which occurred in children under five years of age, are distinguished by this feature. Now, the same symptom is very prominently developed in many lower animals, particularly in the dog and cat, and more especially when a salt of morphia is the preparation employed. In a commentary appended by me to the history of twelve experiments made by myself on the administration of large doses of morphia to dogs, cats, and rats (by injection of an aqueous solution into the peritoneal cavity, an **extremely** regular and accurate method of administration) I find the following remarks: "In the first place, it may be observed that convulsion, in one shape or another, was noted in every case except that of a dog, who received a (for that animal) very small dose. Clonic convulsions were present in every instance where there was convulsion at all. On the whole, it may be stated decidedly, that the violence of the convulsions, and the frequency of their recurrence, was proportionate to the largeness of the dose. . . . . Paralysis, at any rate of sensation, was always developed prior to the occurrence of clonic convulsions." (The quantities of morphia used ranged from one to five grains.) It is with these lower animals, as it is with the young human subject: convulsions are a very frequent symptom of opium-poisoning; and it is important to remark that they always occur simultaneously with paralysis of sensation and usually of motion, and with **profound** unconsciousness. They occur, in short, under circumstances in which it is impossible to imagine any excitation of the medulla oblongata, the great reflective center, for this is too evidently being overwhelmed and paralyzed by the same narcotic influence which has destroyed consciousness by paralyzing the cerebral hemispheres. The same kind of convulsive action has also occurred in adults,[*] and I believe that this happens more frequently as a consequence of poisoning with morphia than with opium itself.[†]

[*] *Vide* Christison on Poisons, p. 707 (Fourth Edit.), for a number of cases.

[†] Ibid., pp. 721–6. Taylor on Poisons, p. 615.

Another important variation, which is occasionally observed in the course of **acute** opium-poisoning, is the greater severity with which the poison depresses the respiration and the heart's action, in proportion to the affection of **consciousness**. Thus, it has been already said, it has not unfrequently happened that the persevering attempt, with larger and larger doses of opium, to procure sleep in delirium tremens, has caused a sudden collapse, with little or no premonitory coma, in which the breathing and the heart's action have rapidly come to a standstill.

There is another course which opium narcosis sometimes takes, which is, indeed, not frequently seen in man, but which is common enough in some of the lower animals. Dr. Sharpey was good enough to inform me that, in experimenting upon *frogs* with an aqueous extract of opium,* he found that *tetanic* convulsions were produced; and I have since repeatedly verified this observation. Moreover, I have observed the same phenomenon, in two cases, in experimenting upon rats with large doses of morphia. Rats are also liable to a very peculiar affection from morphia, which closely resembles the state known as hysteric catalepsy, and which was well shown in the following experiment, as well as in several others: A full-grown rat had three grains of acetate of morphia, dissolved in ʒi of distilled water, injected into its peritoneal cavity. Twenty-seven minutes later the surface of the body was everywhere insensible to any but the strongest impressions. A quarter of an hour later *still* "a partially cataleptic state was developed. If the limbs were placed in any position, however awkward, they retained it steadily. Thus the animal was made to sit upright upon its hind quarters, with its fore legs extended in the attitude of a popular preacher; it maintained this posture for several minutes, the head slouched to one side. On being pushed so as to lose its balance, it recovered itself somewhat, and crouched down." A few min-

---

* There is reason to think that alcohol, in small doses, antagonizes the tendency to tetanic convulsion.

utes later, on being stirred up, the animal walked a little way, lifting the hind legs with great difficulty (they were found to be somewhat rigid). Soon after this the cataleptic condition became more marked, and voluntary movements entirely ceased. Clonic convulsions now set in, and continued to recur frequently for about half an hour. A general tetanic spasm then seized the animal, which lasted for about six or seven minutes, when respiration was found to have ceased: the heart beat faintly for two or three minutes longer. The whole body was quite rigid very soon after death.

In man the symptoms of *disturbed imagination* in acute opium-poisoning are usually little prominent, owing to the rapidity with which consciousness is obliterated by a large dose. That delirious fancies would otherwise form a conspicuous part of the train of phenomena is rendered highly probable from the fact that in the chronic poisoning of excessive opium-eating, in which the toxic effect of the drug is as it were *analyzed* for us, this symptom has been observed to be extraordinarily prominent, as we shall see when we come to deal more fully with the mental phenomena of narcosis. And it is curious to note that, in some animals, especially in the cat, a kind of delirium is produced by large doses of opiates, rapidly thrown into the system, which there is great reason to think is attended with the production of spectral illusions.

Finally, the cat furnishes us with another example of the striking differences which may occur in the action of opium upon different organisms. If there be one symptom which we are apt to think more than another inseparable from opium-poisoning it is *contraction of the pupil*. Yet, in the cat, the action of poisonous doses immediately, and from the first, *dilates* the pupil to a very remarkable extent, and it remains in this condition and totally insensitive to light till death, or till a decided turn has been taken toward recovery.

The above brief sketch of the chief variations which may be observed in the course of the symptoms of acute opium-poisoning may serve to teach us an important lesson. Within the scope of the physiological action of a single narcotic, we have seen examples of a number of toxic effects, which at first sight we should, perhaps, be little inclined to call "narcotic" at all; such, for instance, as the remarkable tetanic convulsions induced, frequently in some animals, and sometimes in man; yet we cannot doubt, when we view such symptoms in connection with all the coincident phenomena, that they are part of the same morbid influence, acting in the same direction, as that which has produced the coma or paralysis which seem more appropriate to narcosis. The whole process is one of extinction of the life of the various parts of the nervous system successively, and is fundamentally different from that operation of small doses of opium in relieving nervous depression which we noticed in the chapters on stimulus. There is a true unity between all the varied effects of large doses. They always indicate poisonous depression of the nervous system; just as there is a real unity between the various medicinal effects of small doses, which always act by restoring some deficient vital action.

The narcotic effects of opium are so various and comprehensive, that they afford us an opportunity of noting what *parts* of the nervous system are indicated as probably injured by the occurrence of such and such symptoms in narcosis. 1. To paralysis of the brain may be ascribed delirium, coma, emotional "excitement," involuntary memory, and involuntary fancy. 2. To paralysis of the spinal cord may be probably ascribed spasms, tetanic convulsions, paralysis of sensation, independent of loss of consciousness, tingling and creeping sensations on the surface, and actual pain. 3. To paralysis of the medulla oblongata may be probably referred clonic convulsions, tremor, and shudderings, disturbances of respiration—*vomiting.* 4. To paralysis of the organic nerves of the heart may be probably ascribed the irregularity or cessation of the co-ordinated movements of that organ. 5. To paralysis of various portions of the vaso-motor system may be referred the abnormal increase and disturbance of the secretions. The arrest of secretion, when this occurs, not improbably depends upon changes in the epithelium of the glands, caused, like increased secretion, by paralysis of the vaso-motor nerves, but unfavorable to secretion in the same way as scarlatina-poisoning and many other forms of "kidney-irritation" are unfavorable.*

Instead of following an anatomico-pathological arrangement, however, I shall prefer, in the ensuing chapter, to trace as clearly as possible a clinical picture of the varieties of acute narcosis, taking the above-mentioned six groups of symptoms successively, and treating them somewhat in detail.

---

* It will be observed that I hazard no opinion as to the effects of paralysis of the cerebellum. The connection of that organ with the function of locomotive co-ordination is at the present day far more doubtful than it formerly appeared to be.

12

CHAPTER V.

Mental Phenomena.

THE mental phenomena of narcosis form one of the most interesting chapters in psychology; and though as yet our acquaintance with them is very limited, enough is known to afford us valuable materials for reflection. In attempting to give some account of them, it will only be possible for me to select a few conspicuous examples of the different modes in which the mind may be disturbed by the action of narcotics, for to describe the infinite variety of gradations between these more conspicuous types would be a hopeless task.

The narcotics which especially affect the brain may exercise their depressing influence in such a way as to render that organ, at once, more or less incapable of ministering to *any* mental operation; or they may lessen, at first, only its capacity for giving effect to certain kinds of mental energy.

The former kind of narcotic influence may be comparatively easily understood. We can readily conceive that a poison, circulating in the blood, and especially attracted to the brain, may pervade the whole mass of nervous tissue in the hemispheres in such a manner as to render these organs more or less unfit for the performance of their function as the immediate instrument of mind. Such a condition is represented by the occurrence of partial or total coma. It is far more difficult to perceive the *rationale* of such phenomena as narcotic inebriation, narcotic delirium, narcotic spectra, and narcotic reminiscence, to which I desire to direct particular attention.

Narcotic inebriation. This remarkable condition has been frequently very incorrectly described. Its first stages have been represented as the result of the preliminary "stimulant" influence which is exerted on the brain by a particular class of narcotics. It has been said that alcohol, the æthers, chloroform, Indian hemp, camphor, and some others, when given in large doses, are at first intensely stimulating, but that a secondary depression quickly arises which disturbs all the intellectual powers. It is, however, highly improper to give the name of "stimulation" to that effect which is produced by a dose of any narcotic which places the taker in such an unnatural mental condition. In order to make this more clearly perceptible, I shall now analyze, with some minuteness, the phenomena of the inebriation produced by alcohol, by chloroform, and by hashish respectively.

When an excessive or narcotic dose of alcohol is taken into the stomach, the symptoms will vary according to the rapidity with which the blood becomes saturated with the poisonous agent. A certain interval will elapse during which only so small a quantity shall have been absorbed through the stomach walls as may stimulate the nervous matter when carried by the circulation to the brain : and this interval will be long or short, according as more or fewer obstacles to absorption have existed, such as dilution of the alcohol, or mechanical obstructions (*e. g.* food in the stomach). But from the moment that a certain degree of saturation of the blood is overpast, and the so-called "phenomena of excitement" commence, stimulation is entirely at an end, as we may plainly perceive, from the fact that the nervous system, in all its parts, is becoming paralyzed. The first warning of alcoholic inebriation is flushing of the face, an occurrence which indicates that the cervical sympathetic is becoming paralyzed; this is a symptom not peculiar to the action of alcohol; but when caused by this narcotic, is always a sign that paralysis of the hemispheres is commencing. It is about this period, or soon afterward, that the drinker finds himself in unnaturally high spirits; that his **animal** passions **are** most prominent; that feelings of vanity carry him away into garrulity of talk, and that whatever sentimentalism there **may be in his nature, is apt to come out,** often **ludicrously enough.** As more and more **alcohol begins** to circulate **through the** nervous system, and especially through the **brain, fresh** mental peculiarities **are** developed, but always *in a certain orderly sequence,* although this fact easily escapes observation. The clew to a right appreciation of the succes- **sive** phenomena is this: that the feelings ordinarily sup- pressed by voluntary effort, or observed by the impressions of **actual life, are displayed, by the** removal of these custom- **ary veils,** *in the order* of *their concealment.* Reason and prudence, **and** the moral sense, which form the varnish, **mostly a** thin one, superimposed upon the sensuous nature, vanish **simultaneously** with the faculty of estimating ideas of time and **space, and** with the power of accurate co-ordination of the muscular movements. It is **now that** the truth of the proverb begins to be most completely realized, "in vino veritas;" for **the very most hidden** things of the mind come out.* Suppressed emotions, **passions,** desires, have their sway. **The** brutal Irish navvie, whose haunting vision of an avenging policeman has vanished into the mist of alcoholic remoteness, pounds his wife to **a** jelly, throws her down, and stamps on her face, after his manner. The foolish "pigeon," **who yet is wise** enough, when sober, to let no one suspect that he believes colossal fortunes are likely to be picked up **over the gaming-table, displays** an eager and genuine confi- dence, when drunk, in his precious system for securing a constant run of luck. And many another ugly phase of human (or shall we call it animal?) nature reveals itself in unsuspected vigor.

The inebriation **produced by chloroform,** in large doses, very nearly resembles that from **alcohol;** the chief differences

---

* So Schiller **says:**

"**Der Wein erfindet nichts, er** schwatzt's nur aus."—*Die Piccolomini.*

between them being due to the greater rapidity with which
consciousness is lost in the former case. Like alcohol, chloroform rapidly attacks the cerebral hemispheres, when administered in large dose and in a form which admits of easy
absorption. Extravagant exhilaration of spirits is usually
the first sign of true chloroform narcosis; this is obviously
produced by obliteration of the consciousness of actual circumstances and surroundings, as is well seen in the case of delicate timorous females about to undergo operation in a crowded
hospital theater. Gradually this exhilaration, which is often
accompanied by a flow of nonsensical talk, subsides, the stage
of inebriation is over, and that of coma, pure or mingled with
delirium, succeeds.

The inebriation produced by hashish differs materially from
that due to alcohol or chloroform. Upon the brain of civilized
men, this narcotic rarely works so as to produce that fierce,
uncontrollable outburst of passion which is so often witnessed
in the drunkenness due to either of the other two agents. Its
effects upon the mind are almost equally decisive, as regards
the obliteration or disturbance of the consciousness of surrounding circumstances, as are those of alcohol or chloroform;
but instead of violent bursts of passion, a placid self-compla-
cent vanity is developed, which makes the subject of it feel
himself the greatest being, physically and mentally, in the
universe. Yet even hashish, taken by the half-savage of
some wild Oriental tribe, has as powerful an influence in
letting loose fierce passions as the rawest whisky has upon
the most brutish navvie. The celebrated Oriental traveler,
De Sacy, believes that the famous "Assassins" derived their
name from the fact that they excited themselves to fury by
the use of hashish in very large doses; and numerous modern
observers confirm the fact of the occasional production of this
kind of effect.

On the whole we may say of inebriation (as illustrated by
the action of alcohol, chloroform, and hashish), that its essence
consists in the destruction of the capacity of the brain for

retaining or recalling moral and prudential impressions, and also for any kind of continuous intellectual labor; and that the apparent *excitement* of the emotions and desires is, in truth, but the unveiling the lower part of our nature, which is more or less ready, in each of us, to spring into action when the customary checks are removed.

The second class of mental phenomena produced by narcosis, which we have to notice, includes those which come under the head of Involuntary Reminiscence. The most conspicuous examples of this are to be observed in the occasional effects of alcohol, chloroform, opium, and especially hashish.

Alcohol, taken in poisonous doses, not unfrequently recalls to the mind circumstances and ideas long since forgotten. This is proved by the talk of some drunkards: the most conspicuous example of it which has come under my notice was in the case of an old gentleman who regularly, every night of his life, intoxicated himself with bottled pale ale. The process of poisoning in his case was a slow one, and it may be doubted if he ever made himself "dead drunk." But he was in the habit, when much too far gone in liquor to walk steadily, of talking to himself in a loud, though indistinct voice, of the history of his life, and expatiating with maudlin gravity upon the enormity of his fault in yielding to the habit of drunkenness, often, as it seemed, without the least consciousness that he was then and there drunk. At other times, on the contrary, he was well enough aware of his actual state, but this was only on occasions when he had indulged with comparative moderation. Whenever he was very drunk indeed, all his sorrow seemed to be that he should have been so intemperate in past years.

Chloroform still more frequently recalls the events and feelings of past life to remembrance. During the short period which elapses between the commencement of an inhalation for surgical purposes, and the arrival of complete unconsciousness, patients often talk of persons and things they have had no connection with lately. The most striking example of this involuntary reminiscence of past things is afforded by the indecent and violent language in which persons of the gravest character occasionally indulge under these circumstances. This painful but happily rare phenomenon can only be accounted for on the supposition that impressions have in past times been received, perfectly passively, by the brain; impressions which have been temporarily effaced by others due to the ordinary course of life and conversation, but which are revived as soon as the poisonous influence has destroyed the power of the brain to minister to the usual modes of thought.

The remarkable power of opium, in some subjects at least, to recall past events to the mind, is strikingly illustrated by De Quincey,* in his chapter on the "Pains of Opium:" and I shall not attempt to repeat a story which I should only mar; merely remarking that this phenomenon is probably confined to a small number of opium-eaters, of peculiar constitution.

It is hashish, however, which produces the most striking effects of this kind. In the experiments which M. de Lucca† made upon himself, with large doses of the sugary paste of hemp, he observed that when narcosis had advanced to a high degree, he "reviewed, with the most intense satisfaction, every action of his life, but his ideas flitted so rapidly through his brain that he was unable to grasp steadily any one thought." O'Shaughnessy‡ relates the case of a youth, of retiring disposition and excellent habits, who, under the influence of a quarter of a grain of the resin of hemp (a powerful preparation), forgot all the immediate circumstances and persons surrounding him, and identified himself with some rajah, of whom he had doubtless previously read or imagined; he displayed unusual confidence and aptitude in conversation, and assumed the manners proper to the potentate he fancied himself, and betrayed the fact of his possessing stores of in-

* Confessions of an Opium-Eater. Author's Edit. of De Quincey's works.
† Journal of Practical Medicine and Surgery, 1862.
‡ Bengal Dispensatory, p. 597.

formation, for which even his intimate acquaintance had not given him credit. It seems to have been a true apocalypse, in which the boy's real character and abilities were revealed.

We can have little doubt, I think, as to the true interpretation of these phenomena of narcotic reminiscence, if we compare them with facts which meet us at every turn in the history of disease, and of mental deficiency or aberration. In a considerable number of maladies, in which the brain is unable to perform its highest functions, and among them that of voluntary, or as we might call it *creative*, recollection, it gives evidence in a remarkable degree of that *passive* memory which we have seen displayed in narcosis. In some forms of insanity and of idiocy, in epileptic delirium, in the delirium of fevers, of chronic alcoholism, &c., this passive reminiscence is sometimes a very marked feature; nor is it less so at the approach of death from various other diseases. Whatever the exact nature of the process may be, it seems to resemble the uncovering of the older inscriptions of a palimpsest by some agency which destroys the later and more superficial writing. The most remarkable instance of this kind, perhaps, which has been recorded is that which Coleridge relates,* of a poor ignorant servant-girl, who could neither read nor write, but who, in the delirium of a fever, repeated entire passages from Latin, Greek, and Hebrew authors, which must have been passively received by her brain years previously, from merely hearing them recited by a learned man in whose service she had lived. A striking case has been recently communicated to me by Dr. Radcliffe, of a gentleman who, having suffered sunstroke in India, entirely lost his knowledge of the Hindustanee tongue (which had been excellent), and became subject to epilepsy: strange to say, this language, which was lost to him as far as the power of volun-

---

* Biographia Literaria, vol. i, p. 117. Coleridge drew a very different inference from this fact to that which is here deduced from it; but that was owing to his erroneous assumption that the delirium of fever implied a hyper-stimulated condition of the brain.

tarily using it went, came freely to his lips in the prostration of epileptic delirium. And still more strange, now that the epileptic tendency has considerably diminished, with returning health his voluntary command of the Hindustanee language is slowly coming back.*

If we had any doubt that these curious phenomena of involuntary reminiscence depend upon the arrest of the higher functions of the brain, it would, I think, be removed by an attentive consideration of those persons in whom these functions have never been performed or have ceased to be performed. It is a common observation that very great calculators, and proficients in *memoria technica*, are frequently below the standard of intelligence in general matters. Nay, more, it has been observed that persons born absolutely or nearly idiotic may possess a surprisingly tenacious memory for certain classes of events. A lady has informed me, from her own personal knowledge, of the case of a boy, " little better than a fool," who nevertheless possessed an extraordinary power of recollection for one class of events, namely, the changes in the weather; so much so, that he could at once and without hesitation say whether it had been fine or stormy on any particular day for years previously. And another boy, known to the same lady, was absolutely an idiot from birth ; yet he had a perfect memory for the history of all the farm animals in the neighborhood, and could tell with unerring precision that this was So-and-so's sheep or pig among any number of other animals of the same kind.

We see in these examples the same thing resulting from conditions habitual to the organism as is apt to be artificially produced by diseases and the action of narcotic poisons. The things which are generally regarded as important—the affairs of actual life—are, as it were, *blotted* from the mind in the latter case; in the former they have never made any impression on it. In either case, impressions of a certain limited

* In connection with this subject, *vide* Sir W. Hamilton's " Lectures on Metaphysics." (Vol. i, chap. xviii.)

class, and of what we may call a coarse and striking kind, which have been received upon the brain, are at liberty, owing to the absence of any interference from the operations of creative intellect, or from anxious speculations about the things of the present time and place, to display themselves in all their freshness. Even in cases of narcosis where, for a time, the intellect seems to work with even more than normal facility (as in the instance quoted from O'Shaughnessy), the prevailing symptoms are still of a paralytic kind, and it is obvious that mental power has not been increased, but only *revealed*.

3. A third class of mental symptoms, produced not unfrequently by narcosis, may be grouped under the denomination of involuntary fancy. At first sight it may appear that this sort of mental disturbance is inseparably bound up, and may be identified with involuntary reminiscence: but this is not the case, although the two kinds of phenomena are not unfrequently coexistent. What we may call involuntary fancy is the *ordinary*, as involuntary reminiscence is the *occasional*, characteristic of delirium; and, on the other hand, I believe that involuntary reminiscence is the usual, and involuntary fancy an extraordinary, characteristic of most non-congenital forms of insanity. Madmen are said to *reason correctly from false premises;* and, as far as my own observation goes, these false premises usually consist of some old impression—whether of an actual event or merely of a strong imagination—revived with preternatural intensity, in consequence of the obliteration of more recent ones by failure of the powers of the brain. The reader will perceive that what I mean by the phrase "involuntary fancy" is the pseudo creative action which presents unreal *spectra* to the eye, or unreal feelings to the senses (*hallucination* of Bucknill and Tuke*), or materially modifies the appearance of natural objects (*illusion* of the same authors). Here the powers of judgment are by no means destroyed, but only temporarily obscured. The impressions made by various

* Manual of Psychological Medicine, Second Edit., p. 134.

objects at some time or other physically presented to the
senses, or mentally pictured upon the brain, recur, not in a
fixed order, so that the patient can connect them with any
particular event in his life, but pell-mell, or at least disorderly, and thus wearing the appearance (though this is
delusive) of being created and not merely reproduced. Of
these phenomena De Quincey gives a most accurate and
graphic description in his chapter, already quoted, on the
" Pains of Opium." He remarks that this part of the effects
of opium-narcosis is similar to a "power" possessed by many
children " of painting, as it were, upon the darkness all sorts
of phantoms. In some that power is simply a mechanic affection of the eye; others have a voluntary or semi-voluntary
power to dismiss or summon such phantoms; or, as a child
once said to me, when I questioned him on this matter, 'I
can tell them to go, and they go; but sometimes they come
when I don't tell them to come.'" This "faculty," as De
Quincey calls it, distressed him exceedingly—for at night, as
he lay awake in bed, " friezes of never-ending stories, that to
his feelings were sad and solemn as stories drawn from times
before Œdipus or Priam, before Tyre or Memphis," passed
before his mental vision, and his dreams also assumed a
similar form. Whatever idea he called up, in the waking
state, by a voluntary act, was very apt to transfer itself to
his dreams; while a constant sense of profound melancholy
brooded over his mind, both in sleeping and in waking
moments. His perceptions of time and space were altered
in a way which caused natural objects, and periods of time,
to be enormously magnified to his imagination. These effects
are not produced in me* by large doses of opium: in my case
a semi-comatose condition is almost at once induced, and the
mind is an utter blank for many hours. (Such was the effect
of a *grain of morphia*, taken by me on one occasion for the
purpose of getting quickly rid, as I thought, of a severe cold.)
But the description of De Quincey exactly agrees with what

* Nor, I believe, in the majority of Englishmen.

I have been told by several opium-eaters; and, moreover, it corresponds closely with the phenomena produced by hashish. Von Bibra* says, when under the influence of this drug he observed that pictures of various kinds presented themselves to his eye, even in daylight, so that he could imagine them painted on the folds of a white handkerchief which he held in his hand; and, by shifting the folds, he could change the pictures. And Von Bibra, De Lucca, and Moreau all concur in the statement that, in hashish-narcotism, an enormous magnification of time and space is one of the most constant symptoms. The only instance of involuntary phantasy from the use of narcotics of which I have personally been the subject was caused by a large dose of extract of belladonna, which I took for the purpose of testing the accuracy of some statements which had been made as to the innocuousness of that drug. One grain and a half of what I believe to have been a very good extract of belladonna was taken on going to bed (11 P.M.). At about 4 A.M. I awoke in a state of slight, but decided, delirium. My judgment, I think, was sound, when I chose to exert it, but nothing could rid my eyes of a legion of most disgusting spectra. I am not very partial to any part of the insect creation, but cockroaches are my special horror; and spectral cockroaches were swarming all over the room. Every object in the room, both real and spectral, had a double, or, at least, a dim outline, owing to the extreme dilatation of the pupil. My hands also shook a little. This state lasted for about two hours, and then passed off, leaving me nothing to complain of but a dry, sore feeling in the throat.

In the year 1857, I was myself the involuntary cause of a nearly fatal case of belladonna poisoning. Owing to the mistake of a nurse, acting under my orders in King's College Hospital, a suppository, containing no less than five grains of extract of belladonna, was allowed to remain all night in the rectum of a female patient, instead of being brought

* Die Narkotischen Genussmittel, &c., p. 281.

away by turpentine injection, as was directed in case it did not discharge itself within half an hour. As may be imagined, the morning found the patient in a most dangerous condition, and life was, with difficulty, saved by the administration of stimulants. At the height of the poisoning, the woman was in a state very closely resembling that so often seen in delirium tremens. Excessive terror was painted on her countenance, and she responded to all questions by pointing with a trembling finger to swarms of unclean beasts, which she fancied were scrambling all over the beds, the walls, table, &c. of the ward. The same kind of *spectra* are usually seen by patients in delirium tremens. The objects which terrify such patients are most frequently of a disgusting character; and, so far as my observation goes, complaints are very rarely made of apparitions which have any reference to the actual circumstances of the drunkard in his ordinary life. In some instances the disgust and terror inspired by such spectra is the cause of acts of violence, and even suicide, by the sufferer from delirium tremens.

In short, the analysis of the phenomena produced by the above-named narcotics, and by others, would lead us to the conclusion that different agents of this class paralyze the brain in different ways, unveiling, in varying order and capricious combination, various antecedent impressions which have been made upon the mind through the brain. That there is anything "creative" about this process, I cannot believe; notwithstanding the opinion of so able an observer as De Quincey. Whatever may be the nature of the dreams which occur in a state of health, a point on which I express no opinion, it appears to me to be demonstrable, in the clearest way, that the delirium of narcosis, and the delirium seen in various forms of disease, is the immediate consequence of the destruction, for the time at any rate, of the brain's capacity for performing some of its higher functions. Such phenomena universally occur in the midst of a general prostration of the powers of the nervous system, and are strictly comparable to the similar phenomena which terrify some children, because the latter have a weak and imperfectly developed nervous system, well adapted for retaining strong impressions, while at the same time they go through but little of the intellectual work, and of the anxiety about the affairs of life, which in later years do so much toward concealing those impressions of an earlier time.

I have dwelt thus fully on the mental phenomena developed by the action of some narcotics, because, although at first sight these may seem a comparatively unimportant part of the physiological action of these substances, they are frequently misinterpreted, and this misinterpretation has a very unfortunate effect in confusing popular ideas on therapeutics. I trust that enough has now been said, although the evidence might be greatly extended, to prove that "stimulation" has nothing to do with the symptoms in question, at any stage. These symptoms are developed in the midst of a spreading paralysis of the nervous system, and, rightly understood, they form an important part of the evidence of this paralyzing process.

Sensory Phenomena.

With regard to the effects on sensation produced in narcosis, there is less difficulty in understanding that these are due to consecutive paralysis of the various portions of the nervous system, than in appreciating the truly paralytic character of the affection which causes some of the mental phenomena which we have just been studying. At least this is true of the effects on sensation produced by the poisonous action of those substances which we are most accustomed to speak of as "narcotics." Diminution of the sensibility of the body to external impressions is well known to be a part of the progressive action of opium, alcohol, chloroform, æther, aconite, prussic acid, henbane, belladonna, Indian hemp, &c., &c. Provided that the course of the poisoning is not too rapid, this diminution of sensibility may be distinguished in the action of *all* narcotics with which I am acquainted, and there are very few that I have not made careful experiments upon with a view to decide this point. The apparent difference, in this respect, between the agents to which the name of *anæsthetic* has been specially applied, and the rest of the narcotic group, appears to me to arise from the order in which the respective portions of the nervous system are affected in the one case and in the other.

If a very careful examination be directed to the progress of sensory paralysis, it will be found, in the great majority of cases, that it commences in the posterior extremities (at least in vertebrated animals), and advances slowly, engaging successively the parts supplied by nerves derived more and more from the anterior portions of the spinal cord. Such is the *general* course of the paralysis, but there is an important exception to this rule in the case of the fifth cranial pair of nerves. It is a curious and at present quite inexplicable fact, that a number of narcotics which produce sensory paralysis, on the whole, in the order above described, produce insensibility of the parts supplied by the fifth pair at a very early period; and some narcotics which do not have this effect on one animal will give rise to it in others.

The main difficulty of tracing accurately the comparative course of the sensory paralysis produced by different narcotics is, that, in the case of those which affect the brain powerfully, unconsciousness arrives so early as to vitiate the investigation, unless the circumstances of the experiment are particularly favorable. The following observations on the action of three narcotics, which in many respects differ in their operation on the nervous system, may perhaps serve to illustrate this point.

Experiment I.—$\frac{1}{24}$-grain of Morson's hydrate of aconitine, dissolved in a few drops of distilled water, was carefully injected, with a Wood's subcutaneous-injection-syringe, into the peritoneal cavity of a healthy kitten (about five weeks old), at 12.30 p. m.

Two minutes later the sensibility was carefully tested, and found apparently perfect in all parts of the body.

At 12.35 the animal appeared somewhat drowsy, but the sensibility seemed still perfect.

At 12.45 the sensibility of the hind legs appeared slightly impaired, but no motor paralysis existed.

At 12.48 sensibility of the hind legs was almost abolished : consciousness was still perfect : a thick slaver was collecting on the animal's mouth, *and the parts about the muzzle were found completely insensible to the prick of a needle.*

At 12.50 vomiting set in, and soon became frequent.

At 1.10 sensibility was found to be completely lost in the posterior limbs, and partially so in the trunk. Motor power much weakened in the hind limbs. Sensibility of the fore legs, shoulders, neck and breast apparently perfect : that of the conjunctiva and the face quite lost.

1.25. Vomiting has apparently ceased : the animal is perfectly paralyzed, as to sensation, in every part of the body, except in the ano-genital region. The fore limbs have also lost all or nearly all voluntary power. The action of the heart is very hurried, irregular, and weak : the respiration sixteen per minute, very gasping.

In this condition the animal remained, with little change, *for four hours*, before unconsciousness was completely developed. Death occurred at 6.30, the respiratory movements ceasing before the arrest of the heart's action. *Rigor mortis* set in immediately, in all the voluntary muscles.

Experiment II.—Five grains of morphia, dissolved in forty minims of distilled water and one drop of acetic acid, were injected into the peritoneal cavity of a full-grown and remarkably strong and lively cat, at 11.46 a. m.

11.49. The animal, which had taken no notice of the operation, and had been purring contentedly, twirled suddenly round several times upon the same spot, with a very excited manner, and with a dazed, stupid look. It now commenced running rapidly round the room, sometimes stopping to sit down on its hind-quarters for a moment.

At 12.5 the animal lay down for a few minutes, panting, and with the heart beating quickly. Sensibility was slightly diminished in the posterior limbs. The tongue was constantly protruded from the mouth, the lips covered with slaver, and quite insensible to impressions. The pupil was very widely dilated and insensible to light. The cat soon got up and began to run round the room again. I now observed that it took its leaps in a very curious manner, seeming to dance on the tips of its toes, and evidently not feeling accurately the surface of the floor.

12.55. The animal has been alternately running round the room and lying down, with its legs folded under its body, the tongue protruded, the mouth covered with foam, breathing rapid, heart's action rapid and feeble. It is now lying down, and there is evidently a great diminution in the sensibility of the fore limbs, and complete abolition of it in the hind legs.

1.10. The animal made a sudden attempt to spring forward, but the limbs gave way—the hind legs being pushed out in a perfectly dead, helpless way, the fore legs extended and struggling. Insensibility of the skin perfect almost everywhere.

About one minute later the cat fell on its side, and went into a series of epileptiform convulsions: it remained profoundly unconscious till death, which occurred at 1.45, by cessation of the respiratory movements, the heart beating some minutes later.

Experiment III.—$\frac{1}{12}$-grain of Morson's hydrate of aconitine, dissolved in a few drops of distilled water, was injected into the peritoneal cavity of a young but strong and lively rat. Respiration was instantly much increased in frequency; and in a few seconds afterward there was shivering, and paralysis, both of motion and sensation, was complete in the

13

hinder limbs : seventy-two seconds from the moment of injection, the whole surface of the trunk and all four limbs were completely paralyzed as to sensation and motion : the animal was still conscious and moved its head—face and conjunctiva still sensitive. The heart's action stopped rather suddenly, at the end of ninety seconds.

Experiment IV.—Ten grains of morphia, dissolved in extremely dilute acetic acid, were injected into the peritoneum of a lively young terrier, at 2.30 P.M.

No symptoms of any kind occurred till the expiration of nearly three hours. (This dog had, three days previously, suffered an injection of two grains of morphia, without its producing the slightest noticeable effect. Dogs are proverbially insensible of the influence of opium.) At the end of that time, it was observed that the hind quarters drooped a little as the animal walked, and on examining them sensibility was found to be considerably diminished. Left alone, the dog sank gradually into the characteristic coma of opium, from which, however, he could be easily roused by shaking him.

At 8 P.M. motor paralysis was complete in the hind limbs, and sensibility was completely lost in the hind, and partly in the fore limbs; not in the face. Shaking the animal still restored his full consciousness.

At 9.58 the dog, which had been lying still, lifted up its head, and began to whine. Sensory paralysis was found to be complete. No impression could make the animal wince at all. Heart's action irregular and intermittent; respiration rather gasping.

At a few minutes past ten, the dog suddenly became unconscious, and at once experienced a violent epileptiform convulsion. For the next twenty-four hours the animal was profoundly unconscious, and the clonic convulsions recurred at very short intervals: the reflex sensibility was much heightened, and it was only necessary to direct a stream of cold air on any portion of the body to induce a fit. At the end of this time consciousness slowly returned, and the animal had completely recovered a few hours later. Paralysis of sensibility and of motion disappeared latest in the posterior limbs.

Experiment V.—ʒiss of whisky, diluted with an equal quantity of water, was drunk by me at 8 A.M., no food having been taken. The pulse was then, and had been for some time, quite steady—64 per minute.

8.15. The face felt hot, and was visibly flushed; pulse 82, full and bounding; slight perspiration on the brow.* A peculiar stiffness and numbness of the lips was noticed, with a slight pricking sensation, not unlike that which is produced by chewing a fragment of aconite root, though less marked.

8.30. The numbness of the lips was more marked, and the perspiration on the forehead was considerable; face very hot and flushed. There was decided giddiness and slight loss of the co-ordinative power in walking, and it was difficult to keep the eyes from converging. No perceptible loss of common sensation in the limbs or trunk; considerable frontal headache.

I now sat perfectly still for half an hour. At the end of this time the symptoms had pretty nearly vanished, slight flushing of the face and headache being the only perceptible relic of the alcoholic narcosis.

During the height of the symptoms my thoughts were confused, and the general impression was that of a nightmare, or, if I may say so, a *day-mare:* there was also slight nausea.

If the action of alcohol be pushed further than this, it will be found that the muscular sense—first of the lower and then of the upper extremities—becomes paralyzed. All this, however, will be more fully described in the Special Researches.

The above experiments are selected from a large number which I have made to illustrate the progressive mode in which

---

* These symptoms are very similar to those noted by Dr. E. Smith. Alcohol *taken in the forenoon*, in ordinary health, rapidly produces narcotic symptoms in a person of temperate habits.

narcotics paralyze the common sensibility of various parts.
The variations in particular symptoms—such, for instance, as
the paralysis of the fifth cranial pair (indicated by numbness
of the lips, &c.)—are very interesting, but there is no time to
dwell on them at present.

The abolition of common sensibility is by no means the only
way in which the deadening influence of narcotics on the sensory nerves manifests itself.   In a considerable number of
cases, *pain* is a more or less prominent symptom, both during
the advance and the retrocession of narcosis.   In some cases
common sensibility is decidedly higher than in health, so that
pain is produced by a comparatively slight impression.

The *rationale* of these latter effects, considered as a part of
the symptoms of narcosis, is, undoubtedly, very puzzling; and,
at first sight, we might naturally feel inclined to deny them
a place in that category.   But a more careful examination of
the facts will show, I think, that there is no just ground for
such denial.

I would beg to recall the reader's attention to the remarks
which have been already made* on the nature of *pain*, in the
argument by which it was sought to prove that the phenomenon is no evidence of increased nerve-action, but rather the
reverse.   I now proceed to consider the significance of an unnaturally *increased common sensibility*, and to endeavor to
prove that, in reality, it is not so directly opposed, in its physiological meaning, to *paralysis of common sensation* as it
appears to our first reflections on the matter.

We have already seen that pain is a phenomenon developed
in a number of diseased conditions which impair the vitality
of parts, whether those parts are or are not known to be supplied with nerves, but more especially in the nerves themselves. We also saw that any circumstance, such as a temporary
increase in the circulation of the painful part, which tends *pro
tanto* to restore it to the normal conditions of vitality, will
usually relieve the pain, unless it be counteracted by some

* *Vide* Chap. II.

hostile influence, such as that of the mechanical structure or the peculiar movements, &c. of the part. When this improvement, however, is brought about by such an event as the occurrence of inflammation in the part, it is mingled with evil, as may be at once conjectured, from the fact that the part, though relieved from spontaneous pain, is more susceptible than before to *painful impressions*. And it is only by a circuitous process, involving, probably, some permanent damage, however slight, to the part, that Nature terminates this kind of cure. How far from the proper standard of vitality the congestive stage of inflammation is, is readily enough seen in cases where the original pain and the present inflammation have engaged some part which contains the apparatus of any *special sense*. Let us take, for instance, the case of superficial whitlow in the finger. Here the spontaneous pain is almost entirely confined to the initial stage, which is marked by shivering and depression. On the occurrence of congestion, and subsequent suppuration, pain vanishes, but excessive tenderness to pressure remains. Now, if the tactile surface of the inflamed finger be carefully examined, and compared with that of the corresponding healthy one, it will be found that there is a marked loss of that *discriminative* power which belongs to the sense of touch, as Dr. Chambers has correctly observed. This is a practical illustration of the law propounded by Sir W. Hamilton, that sensation proper, and perception proper, must indeed always coexist, but that they coexist not in *direct* but in *inverse* ratio to each other.

It is possible, then, for *heightened* common sensibility to external impressions, quite as much as spontaneous pain, and as *loss* of common sensibility, to arise from a condition which is inconsistent with vigorous and healthy life in the part; and this idea is confirmed by finding it noted by Brown-Séquard, that oversensibility to painful impressions was one of the symptoms produced in the parts which lay beyond the point at which section of the sympathetic nerve had been performed in animals. Now, there is a class of poisons—often excluded

by writers from the narcotic group, simply on the ground
that they produce "heightened sensibility" and tetanic convulsion—viz., the strychnia-containing family of plants, and
the *rhus toxicodendron* (or North American Poison-oak). As
to the inferences to be drawn from the occurrence of tetanic
convulsion, we shall have more to say hereafter. At present
we must deal with the import of the so-called "heightened
sensibility." Now, on examination, this proves to be only an
exaggeration of common sensibility, that is of the susceptibility
to painful impressions. Some of the recorded instances of nux
vomica and strychnia poisoning have been distinguished by
the fact that the patients were *painfully* affected by light
and sound. The powers of hearing and of vision, *as discriminative faculties*, are by no means truly increased in such
cases. What happens is simply that the *resisting power* of
the nervous matter which forms the organ of special sense is
diminished by the influence of the poison. It is not that the
patient can see farther or hear better, but that his damaged
optic or auditory apparatus is no longer capable of that "perfect energy," the reflex of which is pleasurable sensation, but
only of an "impeded energy," the reflex of which is pain.
That this is the true explanation of the matter I feel convinced, from my own experiments upon animals with doses
of strychnia insufficient to produce very rapid death, *e.g.* the
following experiment:—A full-grown terrier had $\frac{1}{18}$ of a grain
of strychnia dissolved in twelve drops of distilled water and
three minims of alcohol injected into his peritoneal cavity at
10 A.M.

At 10.15, the animal was found crouching in a corner,
breathing hurriedly; on touching it, it yelped as if in pain,
and trembled all over. On bringing it toward the window,
the eyes were seen to be somewhat protruberant and glisten-
ing, and the dog obviously could not bear the effect of light
upon them.

About seven or eight minutes later, the first symptoms of
muscular spasm came on; the anterior muscles of the neck

and the muscles of the jaw could be felt tightening themselves, with an intermittent, tremulous kind of motion.
Shortly afterward the posterior muscles of the neck were
thrown into an unmistakable tetanic spasm, and from that
moment the symptoms developed themselves uninterruptedly.
Death occurred at 10.35, and from the time that the heightened
sensibility of the skin and of the retina first appeared, they
seemed to continue, to the very end, as far as could be judged
by observation made between the spasms. The animal uttered
no cries except in the height of the spasm, or when the skin
was touched, or the eyes turned to the light. The sensibility
seemed pretty equal at all parts of the surface.

The following cases may also be interesting: 1. M. J. a
man aged forty, a brewer's drayman by occupation, has been
in the habit of drinking *from one and a half to two gallons
of beer per diem* for the last three years, and of taking an
extremely small quantity of solid food. He has become a
teetotaller lately, because he has had one or two attacks of
delirium tremens, but this has not saved him from sinking
into a state of chronic "horrors," and he fears he shall be
driven to commit suicide. Ordered to take strychnia, gr. ₂¼
ter die. Unfortunately this man proved to be one of those
who cannot safely take so large a dose of strychnia as this,
and the following unpleasant symptoms resulted. He came
to me on the third day from his first visit, complaining of
decided stiffness in the lower jaw muscles, a copious flow of
saliva, and "flashes of fire before his eyes." I very particularly investigated this last symptom. So far from his eye-
sight being at all strengthened, it was, by his own confession,
and as tested by myself, much *weakened;* for instance, he
could only read the very largest type, whether at a long or
a short distance from him, and he said that even this looked
dim and misty. The "flashes" complained of always occurred
if he looked toward a bright light; and even in comparative
darkness soon after each dose of the medicine. The sensibility of the skin did not seem notably raised. Of course the

strychnia was at once discontinued; the symptoms thereupon soon subsided. This man's nervous system had been fatally shattered by his long course of alcoholic excess, and he died not long afterward of softening of the brain.

Case 2 was that of a woman aged thirty-three, the mother of a family of six children, and of decent, temperate habits. She was suffering from a peculiar numbness, attended with partial paralysis and some wasting of the muscles of the fore-arm. Strychnia was administered for three weeks, in doses of gr. $\frac{1}{32}$ *ter die*, with the most manifest good effect, both on the arm itself and on the patient's general health. It was then, at the woman's most earnest request, raised first to $\frac{1}{24}$, and then to $\frac{1}{16}$ grain doses; but the use of the latter quantity produced undoubted symptoms of poisoning. The skin of both arms and fore-arms, of the palms of both hands, and of the soles of the feet, became exquisitely tender to the touch so that she could hardly bear to walk, and a considerable tendency to reflex muscular contraction developed itself. But the most distressing symptom was the painful effect which any sound made upon the ear; it was not that sounds were heard at all more distinctly, but that the slightest sound *occupied the woman's whole hearing faculty*, so to speak, and anything like a multiplication of impressions on the auditory nerve became positively painful. The symptoms quickly subsided on the reduction of the strychnia to the former dose of $\frac{1}{32}$ grain *ter die*, and after two or three weeks' longer treatment, the cure was complete.

Case 3.—A young man, aged nineteen, applied to me at Chelsea Dispensary, suffering from incontinence of urine, which had troubled him for three or four years. Having ascertained that there was no discoverable extrinsic source of irritation (*e.g.* ascarides, &c.), and that there was no other indication of special mischief, except the urinary incontinence, I determined to try the effect of strychnia. It was given in $\frac{1}{32}$-grain doses, and gradually increased up to gr. $\frac{1}{12}$ *ter die*, as the smaller doses did not produce a good effect. But the first dose of $\frac{1}{12}$-grain immediately produced impairment of sight and distressing sensibility of the eyes to the dazzling influence of light, and even "flashes," after the eyes had been shut, or sometimes in a dark room. The strychnia was at once withdrawn, and the symptoms quickly subsided. Belladonna, in very small stimulant doses, afterward cured this case.

Case 4.—The patient, before mentioned,* whose case first attracted my attention to the occasional inebriating effects of strychnia in an overdose, suffered from a distressing sensibility to sound during the use of the higher doses, but he was so far from being able to hear plainly, that his first expression in telling me of the symptom was that he was "deafened."

From the description now given of the phenomenon of heightened sensation as caused by strychnia, and by allied poisons, it will be seen that there is little reason for placing these effects in opposition to the phenomena of narcosis. We are quite unable to give any explanation why such different effects on sensation should be produced by different narcotics; but it seems impossible to regard them in any other light than that of different varieties of one species of physiological action upon the nervous system ; an action, namely, which tends to abolish its vital functions.

Muscular Phenomena.

The muscular phenomena produced in narcosis include— clonic convulsions; tremor and shudderings; spasms; vomiting, and other involuntary evacuations; tetanic convulsions : catalepsy; motor paralysis. The consideration of this group of symptoms forms, perhaps, the most important part of our task in studying afresh the nature of narcosis. Even more than in the case of mental phenomena, like inebriation and delirious fancy, it is necessary here to divest ourselves of all

---

* *Vide* Chap. III.

prejudices. The classical doctrines of physiology have taught us to regard the last of the varieties of muscular phenomena above mentioned as opposed in its nature and causation to all the rest. It will be my duty to show that the actual facts of narcosis are to the last degree mysterious and inexplicable upon the basis of these doctrines.

It will not be disputed by any physiologist, that in acute paralysis of voluntary muscles, and still more in acute paralysis of those muscles which are wholly or in part independent of the will, the death (more or less complete) of the nervous system is expressed. It is toward this goal that all the phenomena of narcosis tend,—it is by the complete severance of the channels of communication and co-ordination between the organs, the activity of which is essential to the maintenance even of vegetative life, that the fatal termination is reached. At whatever part of that wonderful circuit of communication which, in all but the lowest animals, seems to be the chief instrument of Vital Individuation, the blow may fall with most fatal force, the last effect is still the same— directly or indirectly, co-ordination of the essential motions is arrested, and death ensues.

Now if once we can bring ourselves fairly to acknowledge this view of the matter, the apparent opposition between paralysis and irregular muscular movement disappears. Given a muscle composed of many distinct fibers, arranged in such and such a position with regard to each other, and as to their physical nature *elastic*, or, if you will, "contractile;" this then we may certainly expect of it : we may expect that if its fibers shorten themselves in concert, an effective or natural contraction of the muscle will result; and on the other hand, that an unequal shortening of the various fibers will produce a knotty or cramped condition of the muscle, and a corresponding want of efficiency in its action. The influence, therefore, be it a compelling or a liberating power, which presides over muscular contraction in the interest of life (that is of individuation), must be a co-ordinative influence. The possession of a machinery of co-ordination may, therefore, be assumed for the lowest organism possessed of fibrous muscles.

But in the case of an organism (like that of the vertebrata especially) in which numerous muscles are placed at considerable distances from each other, the action of which *must*, nevertheless, be combined, it is obvious that free communication, in every direction, and with the utmost rapidity, must be provided for; nay, it even becomes the most important of all provisions for the continuance of life. This purpose is fulfilled by the nervous system, and we cannot, therefore, be wrong in supposing that an agent which breaks through the communications, at any part, is equally to be considered as a devitalizing agent, whatever may be the result to that portion of the muscular system which is thus cut off from the general life of the body. But to say this is in fact to place convulsion, spasm, and paralysis substantially upon an equal footing; and physiological experiment assures us that they are, indeed, closely connected. Palsy of muscles may be said to be the *negative* expression of the fact of their severance from those organs with which they are naturally connected for the purposes of direction, control, and co-ordination. The various kinds of convulsion, spasm, &c. are the *positive* expression on the one hand, of the degree of completeness, the rapidity, the persistence of this interruption of the nervous communications, and, on the other hand, of the varying conditions of the muscle itself, its nourishment, rapidity of circulation, and of tissue change, its electricity, &c. *Rigor mortis* represents the complete destruction of all the bonds of individuation, or "life:" it is the state in which the purely physical qualities of muscle come into play, while as yet putrefaction has not altered the relations of its molecules to each other.

Nowhere do we find a better proof of the justice of this view than in the muscular phenomena of narcosis. Given a slow, equably progressive extinction of the vitality of the

nervous system, as in the operation of poisonous doses of
opium on the majority of human adults, and simple paralysis
will result. **Given a** rapid **extinction, down to a** certain
point, of the vitality of the medulla oblongata, **joined with a**
great **difficulty in** abolishing **it altogether, as in the action of**
large **doses of** morphia administered to **cats, and the most**
violent **epileptiform convulsions ensue. Given the** swift
destruction of **the** life of the spinal **nervous centers** (and
probably of the muscles themselves) **without any very** marked
affection of the brain, as in **poisoning with large doses of**
strychnia, **and we** have tetanic **convulsions, and the** *immediate* occurrence, after death, of the *rigor mortis.*

We may vary the illustrations, almost **infinitely, with the**
same general **result.** Thus the tetanic **spasms which we**
involuntarily associate with **the** idea of strychnia-poisoning,
may be constantly produced in the frog **(an animal in** which
the spinal cord **and** muscles **greatly preponderate over** the
brain in **their development) by the aqueous extract of opium;**
and in several of **the lower animals, and the** *young* **even of**
man, they are a not very unfrequent **result of poisoning with**
morphia. Again, **in the case of the dog, we find an** extraordinary vitality and energy **of the nervous** system, **as we** might
expect from the activity and **the power** of skillful co-ordination of muscular movements which those animals display.
It is almost impossible to overwhelm the great **reflective**
nervous center of these animals with opium, so as to **arrest**
the respiratory movements, and it is by no means easy **to do**
so with chloroform. But with either of these narcotics **it is**
possible so **far to** oppress the medulla oblongata of **dogs**
as to **cause the most** violent epileptiform convulsions. **The**
rat, on the other hand, **is** an animal whose nervous system is
very **rapidly** affected **by narcotics, in such a** way as to produce, first convulsions or sensory paralysis, and then death
by cessation, usually simultaneous, **of the** respiratory movements and the heart's action. For instance, so small a
quantity of the hydrate of aconitine (Morson's) as $\frac{1}{48}$ of a

grain, injected into the peritoneal cavity, repeatedly killed the largest and stoutest rats, in my own experiments, within the space of three to seven minutes. The poison seemed to fall with great violence upon the medulla oblongata, and through the intervention of the pneumogastric nerve upon the heart and lungs, while paralysis of sensation and motion advanced rapidly from the posterior to the anterior portions of the body. Again, death by apnea (cessation of breathing) was caused, in a full-grown rat, within fifty-eight minutes, by the injection of 1½ grain of morphia into his peritoneal cavity. Sensory paralysis, catalepsy, and clonic convulsions were the order of events; simultaneously with a constantly increasing impairment of the force of the respiratory movements, which were at first hurried, and then unrhythmical and gasping. The circulation only appeared to be altered by the fits and by the changes of breathing; it did not appear as if there was any direct powerful impression made upon the heart by the poison.

Catalepsy is one of the least commonly produced symptoms of narcosis, but it is a very prominent one in poisoning with large doses of Indian hemp in man; and as has been already mentioned, it occurs in the rat as a result of poisonous doses of morphia. In neither of these cases does it occur till a certain amount of sensory paralysis has declared itself; but it is one of the earliest muscular phenomena.

Spasmodic rigidity is a symptom not unfrequently produced in narcosis, just previously to the advent of complete paralysis. It is peculiarly prominent in some patients under chloroform; it seems to take the place of epileptiform convulsions, which are rarely seen in chloroformization, in man. The rigidity is often at its height when paralysis suddenly occurs and the limbs drop, perfectly flaccid.

With regard to tremor and shudderings, which sometimes form a part of the phenomena of narcosis, it is to be remarked that they are symptoms not unfrequently observed in patients suffering from epilepsy and other convulsive disorders, and in these cases they bear a similar relation, in time, to the occurrence of positive convulsion, as they do when produced by narcotic poison. I have noticed violent tremor, resembling that of paralysis agitans, in two cases of poisoning with morphia in the human subject, and repeatedly in the course of chloroformization and ætherization, in men and animals; and shudderings are a common premonitory symptom of the tetanic spasm in man and in several animals poisoned with strychnia. Tremor is a very common phenomenon in the period of recovery from acute poisoning with various narcotics; and in chronic narcotism, of which I shall have to speak hereafter, it is one of the most frequent symptoms.

Vomiting, and the involuntary discharge of fæces or urine, hold the same sort of place among the symptoms of narcosis as other convulsive movements. Unless some unusual sensitiveness of the stomach or bowels or bladder exist, or unless they were overloaded previously, involuntary evacuations of them do not occur as an early symptom; their appearance is usually delayed till unequivocal signs of motor and sensory paralysis, and, probably, profound unconsciousness, have been produced. They are not unfrequently among the early symptoms of recovery from narcosis from alcohol, opium, &c., especially in sensitive subjects; in the latter case they are followed by a reaction, which increases the strength of the pulse, and improves the patient's symptoms generally. If, on the other hand, they occur during a stage at which narcosis is still evidently increasing, if the patient still remain pale or livid, and especially if these symptoms be combined with wide dilatation of the pupil, serious danger to life is to be regarded as imminent. This is especially true in chloroform-narcosis in man.

All these convulsive phenomena which we have mentioned are constantly associated with more or less complete paralysis. Motor paralysis usually commences, as does sensory paralysis, in the posterior limbs, and gradually extends itself forward to the anterior part of the body. As convulsion indicates a disturbed condition of the muscle (released only in part from the co-ordinating influence which ordinarily governs its movements), so paralysis represents, we may imagine, the muscle left to itself by the complete cutting off of nervous influence, and as yet maintaining its own equilibrium, as to nutrition, heat, muscular electricity, &c. If the nerves have been completely killed, down to their finest ramifications, we find, after death, that no irritation of a motor nerve, even close to the muscle itself, will throw the latter into co-ordinate contraction.

Such is the rational interpretation, as it seems to me, of the muscular phenomena of narcosis. As in the case of the sensory nervous apparatus, the extinction of life is progressive; and circumstances, which we are as yet unable to understand, determine whether it shall merely issue in paralysis, or whether convulsive or spasmodic movements shall also be produced. Amid all these varieties of effect, one fact stands out clearly; that convulsive movements never occur till such a late stage of the narcosis as necessarily implies that the life of the nervous system is greatly impaired. Even in the case of strychnia-poisoning the apparent increase of common sensibility, which exists between the spasms, is accompanied, as we have already seen, with loss of discriminative power in the organs of special sense. And in the case of most narcotics, motor paralysis has already commenced, at any rate in the hind quarters, before convulsion in any shape occurs.

Secretory Phenomena.

Narcosis produces the most opposite effects upon secretion under different circumstances; sometimes causing a great apparent increase, and at others almost an arrest, of the functions of this or of the other gland. There is great difference between the behavior of different glands; for whereas almost all the narcotics cause greatly increased perspiration from the skin, other glands, as the kidneys for example, pour forth a large quantity of fluid, under some kinds of narcotic poisoning, and little or none in others. The whole question requires far more accurate investigation than has yet been applied to it; but it is not improbable that the rule of narcotic action is, that a certain quantity of the poison rapidly eliminates itself, or seeks to do so, by one or more glands. In the particular gland on which it may throw itself one of two things may happen—either a rapid and copious flow of the watery part of the secretion, carrying with it the poison, or a hasty and profuse shedding of the epithelial cells, such as occurs in the irritation of the kidneys which occurs in many forms of fever, and which is usually associated with diminished secretion.

With regard to morphia, there is little doubt, for instance, that the kidneys are the organs principally destined to bear the responsibility of elimination; and I cannot but suspect that the great apparent insensibility of some animals (*e.g.* the dog) to this poison, may be partly occasioned by the fact of their passing largely increased quantities of urine, and so ridding themselves, in great part, of the noxious agent. In man, however, such increase of the kidney function rarely occurs, in morphia poisoning, as rapidly to set the system free from the evil influence; and it would be very interesting to ascertain whether the state of the kidneys in such cases as prove fatal correspond to the conjecture above hazarded. In alcohol poisoning, the watery element of the urine is constantly much increased, and it has now been demonstrated that a small quantity of the poison passes off in the urine, the breath, and the skin perspiration. Æther and chloroform, on the other hand, are eliminated almost entirely by the pulmonary mucous membrane. The extent to which circumstances may alter the ordinary mode of elimination of a poison, may be judged of from the variations which attend the exit of tartar emetic from the system. This salt, if introduced into the system (by any route) in a time of health, is eliminated by the gastro-intestinal mucous membrane if given in large doses, by the skin and kidneys if given in small doses. But during the existence of inflammatory affections of the lungs, it has been found that large doses of the salt may be given without the production of any vomiting, sweating, or increased urination, the salt eliminating itself through the lungs. The paralyzing action of this narcotic (or sedative, as it is usually called) seems to be chiefly on the sympathetic nervous system, and, to a less degree, upon the spinal cord.

It will be readily imagined that the influence of any particular narcotic upon secretion, the readiness or difficulty with which it becomes eliminated, has a great influence upon its action on the rest of the organism. . We may even anticipate that it will appear hereafter that the so-called "tolerance" which the system exhibits to the action of such an unmitigated poison as tartar emetic in large doses, is to be accounted for solely by the fact of its elimination by an unusual channel; for we find that in chronic antimonial poisoning of a healthy animal (by repeated medium doses), no tolerance whatever is established, and the animal ultimately sinks from the accumulation of the poison, which the kidneys and skin have not been able to eliminate sufficiently.

It is important to note that both the excessive and also the arrested action of glands as a result of the use of narcotics are a true consequence of the *poisonous* action of the latter. They are not at all induced by the ordinary action of such minute doses as only act in a stimulant manner. On the contrary, the result of stimulation, so far as it goes, is to restore the action of the glands to its natural standard: thus the addition of a glass or two of wine daily, or of a small quantity of rum and milk, to the diet of a consumptive patient, often checks the excessive night-sweats which have troubled him, while at the same time it augments his secretion of gastric juice. On the other hand, the use of an excessive quantity of any of the narcotic stimulants infallibly disorders gland-action, producing either excess or deficiency of some secretion or secretions, and always alteration in their constituents, as to quality or quantity, except in the case of such substances as eliminate themselves exclusively or chiefly in vapor by the lungs.

14

It is quite true, however, that the excessive action just spoken of is sometimes used with apparently good medicinal effect. Thus, the ordinary medicinal use of ammonia is as a stimulant, and we have the best reason to suppose that its action in this way is a legitimate one, since it does not produce after-depression. Yet there can be little doubt that we sometimes gain important benefits from its temporary excessive and narcotic action. For ammonia used in full doses is a narcotic, and may even produce fatal narcotic effects (convulsions, palsy of the heart, &c.); and in doses less than this it produces, apparently by its paralyzing influence upon the sympathetic nerve, an excessive action of the sweat-glands which, in peculiar circumstances, may be beneficial. Such an action as this is one of the comparatively limited number of instances which figure in a rational system of therapeutics, *of doing evil* **that good may come**, a practice that should never be admitted in medicine unless the evil is very small and the good very great, and which the advance of our art is constantly teaching us to employ with less and less frequency. It is a practice which is illustrated by the effects already alluded to, of tartar emetic administered in large doses for the cure of inflammation of the lungs, though the good effected in this latter instance is far more problematic to us at this present moment than it seemed to all men's eyes not very many years ago.

Circulatory Phenomena.

The effect of the various narcotic poisons upon the action of the heart and arteries varies greatly according to the dose administered, the patient's health, &c. Ordinarily the first poisonous effects of a narcotic upon the circulation is considerably to increase its frequency and to diminish its strength, and this effect goes on increasing till near the fatal termination, at which time, the weakness still becoming more and more marked, the pulsations become slower and irregular.

But in cases where a large dose is suddenly thrown at once into the circulation, it may happen that the heart's action may be depressed both in force and frequency from the first, as will be hereafter shown to be the case with chloroform inhaled in a concentrated form, and as has been repeatedly witnessed in poisoning with infusion of tobacco, with tartar emetic, &c. In the special researches on chloroform detailed observations will be found on the comparative effects of strong and of weak doses upon the pulse; and the lesson which may be learned from them, and from many similar facts, is important; viz. that very much depends as to the action of narcotics upon the heart, upon the slowness or rapidity with which the circulation is impregnated with a large dose. For it seems impossible to doubt that the action of chloroform, in the strong doses referred to, falls directly upon the organic nerves of the heart and kills them; the organ is found, immediately after death, quite incapable of being stimulated to co-ordinated action.

On the whole it may fairly be said that the affection of the heart usually runs a parallel course to the general paralyzing influence upon the nervous system, provided that the dose is not very large nor very suddenly admitted to the circulation. According to analogy with the other muscular phenomena, we find that the contractions of the heart become more and more feeble and hurried, then irregular and intermittent, and at last cease: that is to say, that the co-ordinative influence of the nervous system is gradually removed from the heart, and its co-ordinated movements are arrested. But that a very strong impression made upon the heart, either indirectly through the brain or directly through the nerves of the heart itself, may arrest all co-ordinated action almost instantaneously.

It is important to remember what was before said as to the necessity of a complicated and perfectly acting nervous system to the co-ordinated movements of the higher animals, in order that we may understand the frequency with which sudden

arrests of circulation are produced in them as compared with
what happens in the case of some inferior animals. It is hard,
for instance, to arrest the action of a frog's heart by chloroform; a very large dose must be introduced into the circulation
by injection, or the heart must be exposed and a stream of
the vapor directed upon it, to accomplish this. The injection
of a weak dose of chloroform will not suffice. But on reflection we can perceive a good reason for this in the greater
tenacity of vegetative life which these animals exhibit. The
heart of a frog will pulsate for a long time after it is severed
from all its connections with the general nervous system, and
even removed from the body.

All differences in the animal organism, and in the mode of
administration of the poisonous agent being allowed for, however, there is no doubt that certain substances have a special
tendency to produce death by cardial syncope, just as certain
others have to arrest gradually the respiratory movements.

Under the head of circulatory phenomena we may notice
also the flushing of the skin, which is an early sign of poisoning with alcohol and with several other narcotics. It is
doubtless due to paralysis of the sympathetic nerves; and it
is interesting as being the first symptom probably (when it
occurs at all) of narcosis.

A more important symptom, or at least one which has usually been considered such, is alteration of the pupil; and if I
venture to include this class of phenomena under the head of
the circulatory changes produced by narcosis, I must be understood to do so with reserve, not as distinctly affirming, but
only believing from the results of modern physiological experiment, that alterations of the pupil depend rather on the
vascular than the muscular system, and are ultimately to be
traced to impressions on the sympathetic nerve.

It may well be questioned whether the common opinion be
a just one, which assumes that contraction and dilatation of
the pupil imply radically opposite physiological conditions.
That they do so *as regards the function of sight* is undeni-

able, for it is certain that the one state is produced by a powerful, and the other by a feeble, impulse of the luminous rays of the sun, or of an artificial light. But in speaking of these two conditions as parts of a general state of the nervous system such as is caused by the action of narcotics, we have no right to assume any such opposition. Remembering what we have said as to the true connection between irregular muscular and vascular movements and paralysis, both being due to interference with the co-ordinative action of the nervous system, we may allow that both the contraction and the dilatation of the pupil which are observed in narcotic poisoning, *may* be due to such interference; and this becomes the more probable when we reflect on the fact that in both conditions the pupil is nearly or quite *immobile*.

That there is anything in the size of the pupil which indicates with any accuracy the *nature* of a particular poison during whose action it is observed, I have long since entirely disbelieved; the circumstance which first led me to this change of ideas being the daily observation of human patients under chloroform. I saw, plainly enough, what I have presented in a tabulated form in the special researches—that contraction of the pupil is an early, and dilatation a late phenomenon of *chloroform*-narcosis; and it occurred to me that it must probably depend on the nature of the nervous system acted on, and the way in which the poison entered the circulation, whether this or any other order of changes in the pupil would present itself; and a more enlarged acquaintance with the effects of narcotics on different animals has tended only to convince me of the justice of this idea. For instance, if there be one drug more than another which has been supposed to exercise a definite influence on the iris, it is opium: the various opiate preparations have always had the reputation of causing *contraction* of the pupil. But in by far the majority of cases of fatal opium-poisoning in men, a closer examination would detect the fact that in the last stage, the contracted state of the pupil is exchanged for one of wide and fixed dilatation, and in many animals the latter effect occurs much earlier than in man. In the cat, for example, the effect of a large dose of opium suddenly thrown into the circulation is invariably to cause wide dilatation and immobility of the pupil *at once* (at least such is the result of my own observations), while the other paralyzing effects on the nervous system are developed in a far more leisurely manner; for instance, the animal retains consciousness for a long time after the dilatation of the pupil—a state of things the very reverse of what happens in adult men, though I have seen it occur more than once in young children. On the other hand, the action which is considered quite characteristic of belladonna—the production of wide and fixed dilatation of the pupil at a comparatively early stage of narcosis in the human subject, varies very greatly in the date of appearance in some other animals, even among those on whose general nervous system the drug has a powerful influence.

On the whole, there is strong reason to believe that fixed dilatation implies the same kind of influence—viz. a paralyzing one—upon the sympathetic nerves, as fixed contraction; although the former indicates that the process has proceeded to a greater length. Dilatation of the pupil coincides, in the case of most narcotics though not of all, with the last stages of narcosis, in which all blood-vessels of the brain are empty (though some venous blood may be found ponded in the sinuses of the dura mater), and at which time also the vessels of the iris would doubtless be illsupplied with blood ; and contracted pupil answers to a stage at which the vaso-motor nerves are as yet only so much paralyzed as to allow a little local congestion of the vessels, among others of the vessels of the iris. It seems to me impossible to believe that an agent like the Calabar bean, which in large doses rapidly paralyzes the heart, can produce the contraction of the pupil, which is so marked a feature in its action, in virtue of anything but a depressing action on the sympathetic nerve. That narcosis may proceed to a fatal termination without the occurrence of dilatation of the pupil at all, proves nothing, so far as I can see, against the theory that contraction is but a minor degree of the same paralysis which issues ultimately in dilatation; since the sympathetic system in such an instance may be the *ultimum moriens*, instead of the *primum moriens*, as is usually the case. At any rate, we **are warranted** in declining to admit the inferences which have been built upon the fact **that** one poison contracts and another dilates the pupil during that stage of narcosis which happens to attract most attention. To imagine, **for instance,** that the poisonous action of belladonna will prove an antidote to the poisonous action of **opium**, upon the general system, simply because **the** former dilates and the latter contracts the pupil in the majority **of** cases in man, appears to me one of the most ill-grounded hopes that was ever entertained by our toxicologists: and certainly the practical success which the plan **meets in** my **hands is** not particularly encouraging. It may be possible that in poisoning with one narcotic, a different narcotic, in doses so small as to have only a stimulant action, might prove beneficial, as is clearly the case with small doses of tea or coffee in poisoning with alcohol. But that the general poisonous action of one narcotic can remedy the general poisonous action of another, appears extremely improbable, and should be rejected till there **is some** better evidence **of it than we at present possess.**

Respiratory Phenomena.

The changes **induced by narcosis in the movements of** breathing **vary a good deal.** In cases where the operation of the poison is slow **and gradual, there is but little if any in**crease of the **breathing rate after the symptoms of narcosis** have commenced; **but the strength of the respiratory move**ments is **progressively impaired, and after a time their fre**quency sinks **below the** normal **standard.** This is particularly the case in poisoning with opium, in the **human adult.** In the great majority of such cases, respiration soon becomes soft and unnaturally slow, sinks into stertor, and at last becomes irregular in rhythm, labored, and gasping, and gradually ceases, the heart continuing to beat for some few moments longer. In other kinds of poisoning, or even where opium has been so administered as to enter the circulation at once in large quantity, there is not unfrequently a considerable increase in the rapidity of breathing; thus in a case of poisoning with three grains of morphia injected per rectum, which fell under my notice, the respirations increased during the earlier stages of poisoning (with strongly contracted pupil, but with delirium and only incomplete coma), the breathing was for some time double its normal frequency; though long before death they sank below the natural standard. In the production of narcosis with æther, and with chloroform (if this be effected skillfully), no change in the frequency of breathing takes place, or only such a change as restores it to its normal standard, so long as the anæsthetic effect is confined to what is technically known as the "third degree;" but as soon as the fourth stage is entered upon, the breathing becomes slower in rate, and stertorous or whiffing in character. Any marked *increase* in the breathing rate (particularly if attended with irregularity of rhythm), occurring after the production of unconsciousness, is abnormal and a sure sign that the process of narcotization is proceeding in an irregular manner. In man this is a comparatively rare occurrence, but in some animals respiration is extremely hurried during rapidly fatal narcotism, until near the close, when it becomes irregular and slow. The effect of rapidly poisonous doses of alcohol on small animals, like the rat, is to produce great rapidity of breathing till death is near at hand, when the respirations become slow and irregular.

Both with regard to respiratory and circulatory changes it is most important to eliminate from the apparent effects of narcotism that increased rapidity which may be occasioned by *alarm*. It is this source of difficulty which particularly besets us in endeavoring to observe the phenomena of chloroform and æther narcosis; but of course the objection does not apply to changes which occur after the arrival of unconsciousness.

Be the respiratory changes what they may, **however, they** all indicate depression of the nervous centers, and that of a very serious kind. For unless in cases where it depends on external irritation, or some other obvious cause, any change in the frequency of respiration which occurs after narcosis has already proceeded to the length of producing unconsciousness to any considerable extent, or marked **sensory** paralysis, &c. indicates that the medulla oblongata is becoming affected by the poison, and that co-ordination of the movements of the respiratory muscles is being interfered with. That both the hurrying and the unnatural slackening of the respiratory movements depend upon a depressing influence **excited** on the nervous system, is shown plainly enough by the action of stimulants, which in the one case calm, and in the other reanimate, **the** breathing. The influence of ammonia, of cold affusion followed by friction, &c. is perceptible even in cases of poisoning with hydrocyanic acid, a narcotic which is especially rapid and destructive in its action on the respiratory nervous centers.

CHAPTER VI.

It is a familiar observation, that narcotic poisons, more than any others, are apt to fail, under certain circumstances, to produce their wonted physiological effect. These circumstances may be divided, for practical purposes, into three classes, according as they concern—the constitutional peculiarities of the patient; his state of health at the time ; and the degree of familiarity he has had with the particular narcotic administered.

Of these modifying circumstances, the first, though highly interesting in itself, may for our present purpose be neglected, as its influence upon general results is trifling in extent, and an investigation of its true nature would be most difficult. The other two require careful consideration.

I. The influence which the state of health of a patient has, in modifying the effects of narcotic poisons, has hardly received the attention which it deserves. There are two instances of this influence which, so far as I know, have never yet been compared, but which will deserve to be considered in connection with each other,—viz. the action of very large quantities of alcohol in fevers, &c. and of very large doses of opium in tetanus. It must be clearly understood that I am not affirming that either of these measures is directly curative. But the extraordinary circumstance to which the reader's attention is now invited is this,—that in both instances the ordinary poisonous action of a powerful narcotic is completely prevented ; for the quantities of alcohol and opium used in the respective cases are often such as would undoubtedly produce fatal narcotic poisoning if taken by the same patient in a time of health ; whereas, in the actual circumstances, no narcotic symptoms whatever appear. In fact the *poison-line* of the narcotic has been shifted by the circumstances of the disease.

Now, both in the case of a fever (such, for instance, as typhus) and in that of tetanus, it is sufficiently obvious that the nervous system, as the great instrument of combination and co-ordination, is rapidly becoming inefficient. What the exact change in it may be, we know not, but it certainly seems no longer as an efficient preserver of that balanced condition of the bodily forces which we call healthy human life. Under these circumstances it is that a narcotic—a substance with a known affinity to that central portion of the nervous system which in the particular case is most severely affected, and which ordinarily exerts on that nerve-center a poisonous influence—loses all its noxious properties, and becomes, at least, harmless, and, as it would seem, most beneficial.

It has been attempted to explain these facts upon the theory that both alcohol and opium, as narcotics, are in their primary action *stimulant*, and that by administering them in repeated small doses the narcotic influence is staved off, and the primary stimulant effect maintained. The fallacy of this argument has been sufficiently exposed in the chapter on the History of the Doctrine of Stimulus. But in truth the theory has not even the most shadowy foundation in fact. The blood is impregnated, not with *small* but with *large* doses of alcohol and of opium, respectively, in the cases referred to, although (for the sake of digestion) the *administration* is affected by small divided quantities often repeated. And large doses of these drugs, when once they have entered the circulation, produce, as we have already seen, nothing but unmitigated depression of the nervous system of healthy persons.

On examining closely the details of these remarkable cases of insensibility to narcotic influence during acute disease, we commonly find that they are distinguished by one fact—namely, the impossibility, from one cause or another, of administering **to the patient a sufficient supply of** ordinary **food.** With regard to alcohol, this will be illustrated fully **in the Special Researches: and the same thing is** evident in the reports of cases of tetanus, in which hundreds of grains of opium have been given in the course of a few days. Such being the case, it is impossible not to be reminded of the fact **that both** opium and alcohol have been frequently used as a substitute for food, even in time of health, when other sustenance could not be obtained, and when the nervous system was especially fatigued; and that under these circumstances it often happened that far larger quantities could be taken without producing poisonous effects, than when the ordinary food-supply was obtained, and no extraordinary fatigue was endured.

Now it is obvious that in such diseases as fever and tetanus, the one necessity, if life is to be saved, is the maintenance of those vital communications of which the nervous system is the channel, and which, owing to its disturbance, are so gravely threatened. But the agents which are used for this purpose are just those which in health would destroy these communications, by lessening the vitality of the nervous system. Nor are we at liberty to suppose that under the new circumstances the alcohol or opium does not act upon the nervous system, since we plainly recognize its influence in the arrest of delirium and tremor in the one case, and of pain and (to a less degree) tetanic convulsion in the other. And these are just the results which we might expect from true stimulant action, such for instance as that of easily digestible and highly nutritious food,* and such as is often exerted by ammonia and other undoubted stimulants. Under these circumstances there would seem to be little difficulty in

* *Vide* Chap. III.

deciding that the ordinary stimulant effect of small doses of alcohol and of opium may be produced in cases of acute exhaustion of the nervous system by doses of these drugs, which under other conditions would be narcotic in their action.

Now to account for the fact that the "poison-line" of one narcotic can be thus shifted while that of another cannot—for example, how it is that enormous doses of alcohol and of opium can be borne without the production of narcosis, in certain exhausted states of the nervous system, while chloroform cannot be taken in larger quantities* than in health—I know not. Extended researches will be necessary before any satisfactory answer can be given to this question; meanwhile, the facts are indisputable, and of the highest importance. The nervous system is obviously in process of devitalization in the diseases in question; its functions are ill performed, and in particular its highest office, that of co-ordination, **is becoming more and more** interfered with. In this state of things it is found that certain narcotics can be given in doses which would usually produce death, not only without producing narcosis, but with a manifestly stimulant effect upon the nervous system; while other narcotics cannot be so given with impunity.

The same facts come out perhaps even more strikingly in cases where a sudden loss of blood has exhausted the nervous system to such a degree as to cause delirium and convulsions, as in severe flooding during or after labor, or in very profuse menorrhagia. In such cases the quantity of alcohol which can be borne, even by delicate women entirely unaccustomed to its use, without the faintest sign of intoxication, is often enormous: I have seen as much as a whole bottle of brandy taken in the course of a couple of hours by such a patient, with the sole effect of restoring the natural action of the heart, and quelling the convulsions, &c. And the quantity of opium which some sufferers from menorrhagia will take

* *Vide* Special Researches.

when suffering from the restless excitement of exhaustion, without a single symptom of opium-narcosis, is surprising. Yet to administer chloroform in a narcotic dose under either of these circumstances, would be highly dangerous, indeed almost **certainly** fatal.

A parallel fact has lately been noticed in the extraordinary insensibility which patients in delirium tremens exhibit to the poisonous action of digitalis. It is true that the whole **subject of the action of** digitalis requires to be reconsidered; still it is impossible to read **the cases recorded** by Christison **and other** high authorities, and more especially **the** experiments on animals by Blake, without coming to the conclusion that this drug is a dangerous poison to healthy persons, in doses such as have been recently administered to sufferers **from** delirium tremens with an effect the very reverse of sedative or depressing. Assuredly, in the latter case, digitalis does not produce its beneficial effects in virtue of any narcotic action.

II. Another powerful modifier of the action of narcotic poisons is the influence of *habit*. This is a difficult subject, and at the same time **of great importance, since** not only physical but moral questions are **bound up in it.**

When we consider the nature of the influence exerted by all narcotics, we can readily understand that the repeated use of those agents must often inflict permanent damage. The very essence of the action of a narcotic is that it paralyzes to a greater or less extent the nervous system (if it be received into the general circulation), and it is difficult to see how it can effect this without causing some change, slighter or more profound, in the physical condition of the nervous tissue. Such a change will demand time to be recovered from completely; and it is easy to see that if the poisonous action be quickly reinduced, recovery must be interfered with. The effect of a number of such disastrous repetitions of the original mischief must at last make recovery, to the full extent, an impossibility.

Such, in theory, would seem to be the necessary course of things: and this is what actually does occur. Physical degeneration of the nervous matter is produced by repeated **excessive** alcohol-drinking, opium-eating, hashish-eating, or **exposure to the influence of lead or** arsenic: for its extreme results are to be seen in softening of the brain, dementia, epilepsy, **motor or sensory paralysis,** &c.; while its more immediate consequences may be traced in mere loss of memory, **muscular tremors, giddiness, impairment of the** special senses, &c. **Such results are, however, by no means constant; on the** contrary, it has been abundantly proved that **an excessive** habitual use of narcotics fails in some individuals **to** perceptibly shorten **life, or** diminish nervous vigor. Some of these instances may doubtless be accounted for by the fact that **the** poison is rapidly eliminated from the system, this is particularly the case with some alcohol-drinkers, in whom **the kidneys are** powerfully affected by their drinks. **In some other cases of habitual excess** with narcotics, a constantly **abnormal** activity of the skin or lungs assists in rapidly removing considerable quantities of the poison from the body.

When neither skin nor kidneys, nor any other channel of excretion is employed in such a way as to prevent the habitual retention of large quantities of the narcotic substance in the **organism, in** the larger number of cases I believe the result **is as follows :—**By degrees the nervous centers, especially **those on which the** particular narcotic used has the most powerful influence, become degraded in structure; this is not merely from the direct repeated action of the poison, **but also from** another cause, viz.—the small amount of common nutriment taken. This is at least the case as regards opium and alcohol, toward which the digestive system seems to have a tolerance, as yet not explained, in virtue of which large quantities of them are at last easily accepted, and have **the** effect of satisfying appetite without causing nausea or **disgust. The habitually** immoderate opium-eater, or alcohol tippler, most commonly takes very little food, but life is supported, in a considerable number of such cases, with little apparent diminution of vigor. The result, however, of this abnormal mode of nutrition is still further disastrous to the nervous system. Deprived of the proper nutriment, which it can only derive from an active supply of blood of the richest and purest quality, the nervous matter tends more and more toward degeneration, and the results of such degeneration are very varied. They may tend to shorten life, or they may not so tend. The changes induced in the nervous matter may be such as may lead to a sudden catastrophe (such as rupture of brain fibers), which may put an end to life at once; or they may consist merely in the gradual shrinking of the brain or spinal cord, or both, in bulk, and the degeneration of a certain amount of their vesicular matter; and this is probably a more frequent issue of chronic narcotism than any positive shortening of life by a sudden overwhelming lesion of the nervous centers. And even where death is premature, in the majority of cases this is due to the general effect on the organism of the state of mal-nutrition which has been **set up, and not to special disease of the** nervous system.

In all cases where degradation of the elements (especially the vesicular) of the nervous centers takes place, it is easy to understand that narcotic effects could not so easily be induced as before. A certain quantity of nervous tissue has in fact ceased to fill the *rôle* of nervous tissue, **and there is** less of impressible matter upon which the **narcotic may** operate, and hence it is that the confirmed drunkard, opium-eater, or *coquero** requires more and more of his accustomed narcotic to produce the intoxication which he delights in. It is necessary now to saturate his blood to a high degree with the poison, and thus to insure an extensive contact of it with the nervous matter, if he is to enjoy once more the

---

* *Coquero* is the appellation given by the Peruvians to those who indulge to excess in the use of the coca.

transition from the realities of life to the dreamland, or the pleasant vacuity of mind, which this or the other form of narcotism has hitherto afforded him.

And herein lies the baser part of narcotic temptation. The genuine debauchee of narcotism *loves to be drunk* with his particular narcotic. He loves to be carried away from all the actual surroundings of life, and placed in a fool's Paradise, filled with illusions of sensual delight. It is quite a different feeling from that of the unwary man who, having taken the narcotic with the intention of using only such a moderate quantity as in fact would not **be narcotic** at all, but would merely relieve weariness, suffers himself to be **persuaded that by increasing** the dose the relief will be **increased.** Such a man as this last has no *desire to be drunk*, which, if I may be believed, who have seen a very great deal of alcoholic intemperance, at least, is *the* secret of the hopelessly downward progress of the ordinary victim of intemperance. **And here** at once we recognize the interference of moral considerations with our physical problem. It is not merely because physical necessity requires a larger quantity of the narcotic to **be taken, that the** confirmed **debauchee** increases **his dose of alcohol, of opium, or of coca,** but it is because his debased moral nature loves the unnatural delights which can only now be obtained by such increase. I confess that this very moral debasement has often appeared to me **more inherent in the** individual's own character than dependent upon the progressive action of the narcotic, mischievous as that may be. It must be remembered that the action of narcotics temporarily reveals the original basis of the character, rather than *inspires* any evil thoughts or feelings. And the attentive study of the mental developments which are brought to our notice under various forms of **narcotism, leads** to the conclusion that there are a considerable number of persons born either with a distinct tendency to sensualism, or with a peculiar susceptibility to certain external impressions, which may remain unnoticed from the time of their

15

passive reception, but **are** always liable to **be** revived. It is among such **persons that the victims of** narcotic excess are chiefly found. There is such a vividness in the impressions which certain ideas of luxurious indulgence, or the force of certain emotions, leave upon the brain of such individuals, that when once these old impressions are revived in their completeness by an agency such as that of a narcotic poison, which obliterates ideas of the present and the actual, their importance and delightfulness is delusively magnified. And henceforward, whatever the general character of the individual may have been, there is more or less of a desire implanted in his mind for that unreal state of feeling which narcotism can produce.

I find myself in danger of treading upon ground which I promised in the introductory chapter to this work to avoid— the ground, namely, of morality. But the reader will perceive that my only object is to define and limit as accurately as possible the physical agency of narcotic poisons; and we are now in a position to understand, generally, what this agency is, in its relation to chronic narcotism. The use of even a single truly narcotic dose very probably produces a real physical damage to the nervous tissue which absolutely requires a certain time for its repair. If the process of recovery be interrupted by an early repetition of the poisonous dose, it will be afterward more difficult, and the reiteration of this vicious sequence will at last render a more or less considerable portion of the nervous system useless as a conducting medium of the peculiar impressions which it is its function **to transmit;** and hence arises an insensibility which **makes larger doses of the narcotic necessary,** as already explained. Moreover, **this insensibility is** accompanied, almost **necessarily, by an habitual feeling of languor and** depression **which is very disagreeable, and with** which the delusions of narcotism contrast very favorably. The dose is repeated, and, for reasons mentioned, in increased quantity; **and the physical** damage to the nervous system progresses in a way which it is not difficult to understand; for although the patient may have brought his nervous system to a state in which the symptoms of narcotic poisoning no longer include pleasant effects upon consciousness, the devitalizing influence continues to be exerted. This is the rule, but there are certain exceptional cases in which, as already remarked, things take a different course. How to account for the fact that certain persons can live, in by far the greatest measure, upon alcohol instead of common food, we as yet know not, but the fact will be shown to be indisputable; and the same thing has been noted with regard to coca, opium, and tobacco, though we nowhere read of such long-continued reliance on these substances in the entire absence of ordinary food, as has been witnessed in the case of alcohol. *In all such instances it is important to remark that narcotic effects are not produced;* and these phenomena must therefore be dismissed from the "narcotic" category.

The explanation of the non-production of the *ordinary* phenomena of narcosis in persons who are habitual abusers of narcotics amounts, probably, to this, that a certain portion of the nervous tissue has been rendered useless for its ordinary purposes.

If we examine carefully the condition of a victim of narcotic excess, in the intervals between his excesses, we shall find that it is a prolonged reproduction of the symptoms which occur during the recovery from even a single poisonous dose of the particular narcotic to which he is addicted. The system of such a person rarely, if ever, contains less than a poisonous quantity of the narcotic which he employs, and hence his nervous tissues have never the opportunity fully to recover themselves, but remain, even at the best of times, at a point intermediate between health and well-pronounced narcosis.

The process of recovery from the poisonous influence of a narcotic, in a person not addicted to its use, is well worthy our consideration. Speaking generally, it may be said to be a reversal of the phenomena of the induction of narcosis; the parts of the nervous system earliest affected being the last to lose the influence of the poison. But in the first place it is to be noted that the brain (in the case of those narcotics which prominently affect this organ) loses the influence of the poison far more quickly than it should do, according to this rule; and secondly, as a consequence of the restoration of consciousness which this brings about, the mind takes cognizance of the condition of the still half devitalized sensory nerves, and there is therefore usually the sensation of neuralgic pain; thirdly, instead of active clonic or tetanic convulsion, which are never present, we have very frequently, indeed almost constantly, *muscular tremor:* and vomiting is an extremely common symptom. Now, these are just the symptoms which are so common in chronic narcotism of different kinds. The morning vomiting, the giddiness, and the muscular tremor which are almost constantly among the phenomena of recovery from acute alcohol-narcosis, are also among the constant phenomena of chronic alcoholism. The headache, stupidity, dyspepsia, and depression, which are the ordinary after consequences of an overdose of opium to one unaccustomed to its use, are among the torments of the professed opium-eater, in the intervals of his debauches. And the very similar symptoms which trouble the victim of habitual oversmoking are likewise closely allied to those which distinguish the period of recovery from the depression occasioned by one large overdose of smoke taken by a novice.

A kind of chronic narcotism, the very existence of which is usually ignored, but which is in truth well marked and easy to identify, is that occasioned by habitual excess in tea and coffee. There are many points of difference in the action of these two substances, taken in poisonous excess; but one common feature is very constant, viz. the production of muscular tremor. Out of twenty-four cases of undoubted excess in the use of tea, and an equal number of instances of extravagant consumption of coffee, which I have collected, there were only five patients who did not exhibit this symptom, which I am inclined to place to the score of the theine which is contained in both these drinks.   The paralyzing influence of narcotic doses of tea is further displayed by the production of a particularly obstinate kind of dyspepsia; while coffee disorders the action of the heart to a distressing degree. I believe that a very much larger amount of illness is caused by intemperate indulgence in these narcotics than is commonly supposed.

An important circumstance, which ought to be remembered, is the difference which exists between the action of these narcotics which are, and of those which are not, easily eliminated from the body.   Of the former class chloroform is the most typical example; if the dose taken be not fatal, the whole, or at least all but an inconsiderable quantity of this poison, insufficient to cause serious disturbance, is eliminated within a very short time, in the form of vapor, with the breath.   And correspondently with this peculiarity we find, as will be fully shown in the Special Researches, that the effects of chloroform are far less permanent than those of alcohol, and still less than those of opium; for the same narcotic dose may be daily repeated, for an indefinite length of time, without any considerable diminution in the effect produced.   Whereas nothing is more familiarly known than the absolute necessity, after a slightly prolonged use of opium in narcotic doses, to increase the quantity given, if we would produce its former effects.   From these facts it would appear probable that some prolonged sojourn of the narcotic poison in contact with the nervous matter is necessary in order to produce that condition in which the nervous system becomes partly insensible to the action of any but an increased dose.

CHAPTER VII.

WE are now enabled to sum up our information on the subject of the present chapter into certain conclusions:

I. Of the numerous narcotics, of most different kinds, whose action we have examined, all, as it would appear, are distinguished by one central feature which marks their influence upon the organism, viz. the production of consecutive paralysis of the various portions of the nervous system.

II. These effects do not appear to be caused, in the slightest degree, by doses which are of less than a certain amount.

III. The dose of any narcotic necessary to produce narcosis, varies according to the age, constitution, and state of health of the taker.

IV. Most, if not all, narcotics when given in smaller doses than those required to produce the lesser degrees of narcosis, act as stimulants.

V. This stimulant action is not produced even in the slightest degree by narcotic doses.

VI. There is, however, a delusive appearance of the mixture of stimulant with narcotic effects, which is caused by the fact that it is impossible, in most cases, for the whole of the narcotic dose to enter the circulation at once. So long as only a stimulant dose has entered the blood, stimulant effects will be produced; but from the moment that the blood is impregnated with a certain proportion of the agent, it behaves as a narcotic, that is a paralyzer of the nervous system. As for those symptoms (*e.g.* excessive secretion, irregular muscular action, delirium, emotional excitement, hurried circulation, and breathing, &c.) which have been spoken of as caused by stimulation, they are in fact the result of a physiological state which it is the very business of stimulants to rectify. For

VII. True stimulation is the supply of some missing influence requisite to the maintenance of that balance of the powers and materials of existence, which we call life.

VIII. The type of the stimulant class is therefore found in foods, with their power of adding to formed tissue, or to the unformed elements of the body (*e.g.* water, salts, or metals deposited or held in solution, &c.), and by evolving heat (by their chemical transformation), which again may be converted into electricity, &c.

IX. Of the medicinal stimulants some are actually known to possess these properties, and others may be strongly suspected to possess them. Thus phosphorus, a powerful stimulant in depressing disease, is a most important element of nervous tissue. Many of the volatile oils, besides their hydrocarbonaceous materials, contain sulphur—a very important element of the body. Between such substances as these, and the common foods of every-day life, we might place such medicines as cod-liver oil, iron, common salt, and the like, which very obviously form part of the structure of the solids or the liquids of the body. It may be preferable to reserve the name of stimulants for the more rapidly acting remedies; but this is the only distinction which we would desire to see made, between "foods" and such stimulants as we have now mentioned, *in a therapeutic point of view.*

X. Finally, there are a number of substances of which as yet we are not able to prove that they are either used for the repair of the tissues, or transformed in the body so as to generate heat; in this class we place alcohol, chloroform, the æthers, various alkaloids, *e.g.* strychnia, morphia, hyoscyamia, atropia, daturia, &c. &c., and the vegetables which contain them: these agents produce, *in small doses*, stimulant effects which are analogous in their main features to those of the remedies, already mentioned, which can claim a nearer relationship to "food" as that word is popularly understood.

XI. All these three classes of stimulants produce effects which, for therapeutic purposes, are similar in kind. They restore flagging vital actions, and check excessive or irregular movements—muscular, secretory, circulatory, &c.

XII. In no case is their stimulant action followed by any other "recoil" than that which is implied in the gradual cessation of their action, involving the necessity of repeating the dose, or not, according to circumstances.

XIII. What has been commonly spoken of as the "recoil" from the stimulant action of a true narcotic is, in fact, simply the advent of narcosis owing to a large, after the occurrence of stimulation owing to a small, impregnation of the blood with the agent. Thus a man drinking 4 ozs. or 6 ozs. of brandy gradually, has not in reality taken a truly narcotic dose till perhaps half the evening has worn away; previously to that he has not been "indulging in narcotism" at all; nor, had he stopped then, would any after depression have followed, for he might have taken no more than 2 ozs. of brandy, equal perhaps to 1 oz. of alcohol. But he chose to swallow the extra 2 ozs. or 4 ozs., thus impregnating his blood with a narcotic mixture capable of acting upon nervous tissue so as to render it incapable of performing its proper functions. The narcosis has no relation to the stimulation, but one of *accidental sequence*. This is proved by the fact that in cases where a narcotic dose is absorbed with great rapidity no signs of preliminary stimulation occur.

CHAPTER VIII.

It seems scarcely necessary to say that there is no sort of connection between the *acrid* poisonous action which certain of the narcotics exercise, in addition to their narcotic power, and the stimulation which may be produced by the administration of small doses of most of these substances. It is as well, however, to clear up this point distinctly.

It will be found, on examination, that the irritant qualities which have caused this particular class of narcotic poisons to be treated apart by systematic writers, as separate from others, are invariably connected only with the action of large doses. But it must be noted further, that a great deal of confusion has been introduced by the carelessness with which the term "irritative" has been assigned to symptoms (*e.g.* vomiting independent of any poison in the stomach) which are really a direct consequence of narcotic depression of the nervous centers.

The kind of irritation to which I refer is different to this; and consists in positive inflammatory affection of the mucous membrane of the alimentary canal or genito-urinary tract. And it must be understood not to include the irritation caused by the *local* contact of certain poisons which, had they been absorbed, would have acted as narcotics upon the nervous system: *e.g.* as when concentrated alcohol, prussic acid, or æther, remains for some time in contact with the walls of the stomach, in which case they act in the manner of pure local irritants. What is meant, or should be meant, by "narcotic-irritant" action is the simultaneous production of narcosis, and of inflammation of mucous membranes, by an absorbed and circulating poison.

Limited in this way, however, the class of acro-narcotics begins to have a somewhat indefinite character. It contains, nominally, the whole of the narcotic group, with the exception of opium and its salts, hydrocyanic acid, and the narcotic gases;* but the production of irritant effects by any of the long list of toxic agents thus classed together is always a matter of uncertainty, and in the case of a large number of them is an infrequent occurrence. An immense field lies open for careful investigation as to the possible share which paralysis of the vaso-motor nerves, more or less complete, and itself a part of the narcotic influence, may have in producing these symptoms, when they do occur. The action of turpentine is a very strong motive to such speculations; this narcotic produces an irritant effect on the genito-urinary tract which strongly wears the appearance of originating in a powerfully depressing action of the poison upon the vaso-motor nerves of the kidneys, through which it seeks to eliminate itself.

Be this as it may, however, it is at least certain that the irritation we are speaking of never occurs except as an advanced symptom, as a consequence of a large dose of the poison, and during a state of general nervous prostration. It is by no means entitled to be considered a result of stimulant action: on the contrary, whatever causes it must certainly be of an intensely depressing nature. And this is the point which it was the purpose of the present chapter to elucidate.

* I am referring to Dr. Guy's arrangement, certainly the clearest and best, where some amount of confusion was unavoidable in the present state of our knowledge.

CHAPTER IX.

GENERAL CONCLUSIONS.

It now remains to combine the several threads of argument which have extended through the foregoing pages so as to afford a view of the general results which flow from them. And here we shall reverse the order in which the phenomena presented to our notice have been examined, and consider, first of all, the idea of narcosis which has been evolved.

The effect of a comparative examination of the various phenomena produced by the action of different kinds of narcotics, is to develop a striking fundamental unity in their modes of operation. It has been seen that agents, apparently so widely separated in many notable external features of their operation, as morphia and strychnia even, in reality follow essentially the same course in the influence they exert upon the nervous system. Upon this portion of the organism, narcotics of every kind exert a continuously devitalizing power.

Once more it is necessary to insist upon the supreme importance of the natural family of symptoms, so to speak, which are seen to be thus closely bound up together. In order fully to understand this importance, it is necessary to appreciate justly the *rôle* of the nervous system in the life of the higher classes of animals, and especially in man.

Life is, for us, so mysterious a thing that it eludes all absolute definition, and can only be fitly described by the use of a word which expresses some unvarying and peculiar condition of its existence. We have seen that the only such condition which it is in our power to fix upon is—Individuation. But individuation implies, of necessity, the most inti-

mate and constant relation between the parts united in the
individual whole; and this is obviously **its** first requisite.
Accordingly, in proportion as we ascend through the scale of
organization, from the simpler forms to **the more complex, the**
demand for a special medium of *communication* increases in
urgency. At a comparatively early stage **in our ascent we**
find one such medium—the circulation of the **blood, developed**
with a perfection of arrangement which secures *completeness,*
at least, of connection.   But, in the highest animals, and
especially in man, we find a type of organization which implies the most complete subordination of parts to the whole,
by means of their complete sympathy with each other: the
mechanism by which this is effected is a highly complex
nervous system.

And here we may refer once more to the researches which
seem to be preparing the way for a more accurate knowledge
than we yet possess of the nature of those communications of
which the nervous system is the conduit. It neither has
been, nor is likely to be, proved that electricity is the sole
dynamic agent in nervous function, but it is now impossible
to doubt that electric changes make up a most important part
of that process. And as I have already adverted to the
inferences drawn by Dr. Radcliffe from the general body
of experimental research on this subject, it is now necessary to observe that the opinions of that author have undergone a modification (since the chapter was written in which
I described those inferences), which must here be briefly
explained.

The basis of Dr. Radcliffe's argument remains the same.
For instance, the bearing of all the observed facts indicates,
in his judgment, that the removal of natural electricity from the
body, rather than its presence there, is the cause of muscular
contraction. But, whereas formerly he was inclined to explain the phenomena by a reference to changes in the *currents,*
which have been shown to exist in every portion of the ani-

mal organism, he now* suggests that these currents are probably but secondary ; and that the essential primary condition of living muscle and nerve is rather one of *static* than of current electricity. It is the disturbance of the natural static electrical condition which may be supposed to give rise to the phenomena of contraction.

The chain of argument runs, briefly, as follows :—In living muscle and nerve, evidence can be obtained that the positive and negative electricity are arranged in a certain definite and constant relation to each other, during the state of inaction. Namely, that the interior portions of the fibers are negatively, the coating positively, electrified. On the near approach of rigor mortis, it is found that these relations are reversed, and the interior of the fibers is positively, the coating negatively, electrified. On the actual arrival of rigor mortis natural electricity is no longer to be detected. Now it is observed that the second condition (that of reversed electrical relations of the parts of the fibers) presents itself in the case of various injuries; thus, in an experiment of Du Bois-Reymond, the nerve of an amputated frog's leg being included in the circuit of a galvanometer, by approaching a hot iron to the nerve the operator can cause an immediate swing of the needle many degrees to the other side of zero on the index ; and, moreover, if the injury inflicted be repaired, by allowing the nerve to regain its lost moisture, the needle slowly regains a position which indicates the restoration of the natural electric relations.

In the next place, it may be observed that, during the state of inaction, living animal tissues furnish evidence of a natural electricity of *high tension:* the proof of this is found in the experiments of Gardini and Hanmer, of Ahrens, Nasse and Radcliffe, with the electroscope: the remarkable discovery by Humboldt of the possibility of an electric discharge across a considerable atmospheric interval between two ends of nerve

* *Vide* Lectures delivered at the Royal Coll. of Physicians. *Lancet*, 1863, February to July.

which are included in a circuit partly animal and partly metallic : the researches of Aldini : and last, not least, the discovery of the *muscular pile*, by M. Matteucci, which possesses the power of decomposing iodide of potassium, and gives signs of tension with a delicate condenser.

According to Dr. Radcliffe,* we are to look upon living muscular and nerve fibers as composed of a number of electromotive molecules, which, from the arrangement of their positive and negative electricity, must, during inaction, always be in such a state that the molecules of the interior are *mutually repulsive*—the molecules of the core in the same condition, and the molecules of core and coating mutually attractive. This being supposed, it is easy to perceive that in muscle the maintenance of the living static conditions would imply a state of *elongation of the fiber*—for the molecules of the fiber would tend to separate from each other in the direction of length, and to approach each other in the direction of breadth. Nay more, from Coulomb's experiments on the electric repulsion of the molecules of elongated cylindrical bodies, it is probable that even were the core and coating similarly electrified, repulsion would be strongest in the direction of length. And the converse of this latter idea leads, without difficulty, to an explanation of the diminution in length, and increase in breadth, of the fibers when natural electricity has disappeared, as in rigor mortis.

With regard to the state of living muscle and nerve, *during muscular action*, Dr. Radcliffe considers that the important observation of Matteucci, that contraction is attended with an electric *discharge* similar to that of the torpedo, is supplemented by the observations made by Du Bois-Reymond, and also by himself, in such a way as to show that during the action of the muscle natural electricity, to a great extent, *disappears* from muscle and from nerve. It would seem, then, that ordinary muscular contraction and rigor may both depend

* For the argument by which this modification of Du Bois-Reymond's "peripolar molecules" is justified, I must refer the reader to Dr. Radcliffe's Lectures *loc. cit.*

upon the absence of the natural electricity of *inactive* living muscle and nerve; and this conclusion is fortified by the fact that, in animals poisoned with strychnia, killed by repeated powerful electric shocks, or dying of certain diseases accompanied with intense nervous exhaustion, there is no appreciable interval between death and the occurrence of rigor mortis.

Of the facts observed in connection with the effects produced by the application of *artificial* electricity, the most striking is this—that the action of *positive* electricity upon living nerve is to suspend muscular contraction and to reinforce irritability—that of negative electricity to increase muscular contraction and diminish irritability. But it would appear that the former must reinforce the natural electricity of nerve, by increasing the positive electricity of the coating, and thus (by induction) augmenting the negative electricity of the core; while the latter must diminish the natural electricity of the nerve, by neutralizing a certain portion of the positive electricity of the coating, and thus (by induction) diminishing the negative electricity of the core. The belief that positive electricity diminishes contraction, appears to be supported by recent observations on the effects of machine electricity upon patients suffering from various spasmodic affections, some of which I have witnessed. The explanation suggested by Dr. Radcliffe of the respective action of positive and negative electricity, appears to harmonize with the fact, which is well established, that artificial electricity can only produce contractions so long as a certain amount of natural electricity remains.

Dr. Radcliffe continues to regard *pain* as upon a level, in vital significance, with irregular muscular action.

It is not my business to pronounce a positive opinion upon the merits of the theory thus propounded by Dr. Radcliffe. But with regard to the main (that is the practical) point— the discussion, namely, of the vital conditions favorable to muscular contraction—I may be allowed to say that the evidence gathering on all sides presents an appearance which is strongly in favor of the idea that the essence of this physiological process is physical rather than *specially vital.* I must have stated my case very ill if the same lesson has not been derived by my readers from the analysis which I have given **of the phenomena of narcosis. Everywhere in the** field thus displayed we see the same sequence of things—the progressive destruction of the living powers of the great instrument of co-ordination—the instrument, that is, of the special individuation, which is characteristic of the highest Animal Life. It seems impossible to separate the various convulsive affections due to narcosis from the category in which they so obviously stand. Nor can we, I think, overlook the strong indications of a regular progress, in degree, from the slighter to the more grave interruptions of co-ordination. From the slightest tremor of chronic alcoholism to the most severe tetanic spasms of acute strychnia poisoning, there is not a single stage of the descent at which we do not find ourselves confronted with simultaneous evidences of paralysis, which, when once looked for, cannot **be missed. Every variety** of irregular and untimely muscular contraction, in narcosis, is connected with the severance of the lines of vital communication : since every such variety is seen, in turn, to accompany the more or less complete severance of those lines by the paralyzing influences which produce the narcotic state.

Such being the case, future researches on the electric conditions present during narcosis may be expected, and will probably elicit most important facts.

Meanwhile we are already in a position to gather our ideas of narcosis into one distinct and intelligible form, and to separate its phenomena from others with which they have been improperly confounded. They stand out as a sharply-defined group, from the indications which they all convey of progressive destruction of nervous life by a poisoned blood supply. What may be the ultimate result of the researches which are going on as to the part which the blood takes in this process

is uncertain ; but it is a most significant fact that Dr. Harley*
has discovered that narcotics so opposed to each other in
apparent characters as strychnia, hydrocyanic acid, æther,
and quinine in large quantities, have one and the same influence, viz. a lowering one, upon the rate of oxidation of the
blood.

In short, every consideration points to the conclusion that
in the work of narcosis stimulation has not the slightest share ;
and we thus establish a radical difference, not merely one of
degree, between the action of small and of large doses of the
substances which can produce both effects.

Separated thus completely from narcotic action, with which
it has no relation save an accidental one, it has been seen that
stimulation resolves itself into a group of physiological effects
of which we give the simplest expression in saying that they
one and all tend to restore the healthy standard of some vital
process or processes. The final step in the definition of stimulants, and that which is doubtless the most open to question,
is the comparison which we have instituted between their
action and that of foods.

In likening the effects of stimulants to those of foods, I
need hardly say that I meditate no attack on the principles
which common sense, no less than physiology, has established
as to the need of the healthy body, in the common circumstances of life, for the supply of certain elements of tissue,
and certain materials for combustion and the generation of
heat. Nor do I concern myself at this stage with the question
whether certain stimulants do or do not possess one or other
of these qualities. It is on different grounds that I would
ask for a revision of the current definitions of the word food.

I think this revision may be demanded on the ground of
the obvious confusion and uncertainty which exists on this
point, along with a great deal of very positive assertion in
physiological works. I can hardly illustrate this better than

* "The Chemistry of Respiration." *Brit. and For. Med.-Chirurg. Review*, October, 1856.

16

by quoting the physiological definition of food given by Béclard, in his splendid work recently published.* " A food," says this author, " is a substance which, introduced into the digestive apparatus, can furnish the elements of the repair of our tissues, and the materials of animal heat." Further on he gives us a definition which, as he says, " is less general. Every alimentary substance, in order to penetrate the organism, introduces itself by the channel of the blood, either directly through the vena portæ, or indirectly by the lacteal vessels and the subclavian vein. Food must, therefore, form a constituent part of the blood itself for a longer or shorter time. We will say, then—*every substance identical with one of the principles of the blood, or capable of being transformed into one of those principles by digestion, is a food.*" Let us examine these two definitions, separately and in comparison with each other.

With regard to the first, it will be observed that it excludes from the list of foods all substances which do not possess the double property of forming tissue and generating heat. It thus excludes water, for though that substance may be said to assist in forming tissue (inasmuch as it is an integral part of every tissue of every living animal organism), it certainly does not become chemically transformed so as to generate heat in the body ; on the contrary, it passes out unchanged, after fulfilling its purposes, in the various secretions and in the air expired from the lungs.

With regard to the second definition, it is to be remarked that it only escapes the anomaly involved in the first (of excluding water from the list of foods) by adopting a different and irreconcilable view of the nature of alimentation. For whereas the former dictum rejected from the list of foods everything that was not, in itself, an epitome of *all* the requisites of nutriment, that which we are now considering admits to that rank anything which is, or can be made, an element, however humble, of the blood. But the ground on

---

* Traité Elémentaire de la Physiologie Humaine. Par J Béclard. Paris, 1862.

which this new definition is professedly based is a very strange one. Briefly it is this :—All foods enter the circulation. Food must, then, form a constituent part of the blood for a longer or shorter time. *Therefore*, every substance identical with one of the principles of the blood, or capable of being made so by digestion, is a food. So little connection is there between the premises and the conclusion, in this argument, that it would be almost as logical were it to run thus :—All poisons enter the circulation. Poisons must, then, form a constituent part of the blood for a longer or shorter time. *Therefore*, every substance identical with one of the principles of the blood, or capable of being made so by digestion, is a poison. I mention these objections to the propositions of Béclard to show that even the most eminent physiologists are liable to fall into a trap when they attempt to define the nature of food **by reference solely to** physiological chemistry.

Surely the only real test of the alimentary character of any substance, is its power to support life for **a** longer **period** than it could subsist if deprived of all external help. **Let us** see how this principle would work. The most important **food** would then be oxygen; for withdrawal of this substance renders the maintenance of life impossible after a **few** moments; while a supply of oxygen, diluted with what most physiologists hold to be merely an indifferent substance (the nitrogen of the atmosphere) is itself sufficient to maintain life for many hours. **Next to oxygen we** should rank Water, without which large mammalian animals perish within three or four days, but with a supply of which, **in** addition to the atmosphere, the want of other food may be sustained for many days, without the occurrence of fatal inanition. Next after these in importance would come the substances more familiarly known as foods; the nitrogenous matters, the hydrates of carbon, and the hydro-carbons which compose the bulk of our ordinary nutriment; the withdrawal of either of these classes of food will injuriously affect the health, and after a longer or shorter period occasion death—the production of the fatal result being greatly slower than in the case of the withdrawal of oxygen or of water. And, lastly we should reckon the inorganic matters—the salts, iron, &c. which are, indeed, absolutely **essential** to healthy nutrition, **but** the withholding of which only very gradually **produces serious or fatal** results.

Such are the material agencies of alimentation; but they are not all that is necessary to this process. Certain external **dynamic influences, and particularly Heat,** Light, and Electricity, are inseparably mixed up with the work of vital sustentation. **The withdrawal** of any considerable amount of this normal heat, for instance, from the media by which the body is surrounded, is fatal to life with a rapidity only inferior to that with which privation of oxygen brings about **the** catastrophe. **It** might, **therefore,** be objected that it **would** be as reasonable to speak of heat as a food as to bestow **that title upon oxygen.** My only answer to this would **be to** allege the convenience of separate names **for** the material and the dynamic supports **of life, respectively.**

Now, if we turn from alimentation **in health to sustenance** in disease, particularly **in diseases** which materially hinder digestion and assimilation, we perceive great changes in the **above scale. There is** the same pressing need **of** oxygen and **of water.** But the comparative value of other articles is **much changed.** Thus, in acute rheumatism, we find the **organism** supporting very well the absence for some days of **the** ordinary nitrogenous **diet, which even** seems to have **become** hurtful, while, on the other hand, the diffusible hydrocarbons, particularly alcohol, are often surprisingly useful. **Thus,** in acute typhoid fever, we frequently find it useful to administer doses of hydrochloric acid, which in health would infallibly **cause** serious derangement, and with this treatment alone it is occasionally possible to support the organism, even without, or almost without any ordinary food, or any alcohol. Thus, in typhus, in low pneumonia, and in pericarditis, alcohol **has not** unfrequently proved able singly to sustain life for **many days.** Thus, in certain forms of peritonitis, turpentine has been observed to support the vital powers in an extraordinary manner. The most remarkable part of the action of these substances is their so-called "toleration" by the system: the ordinary symptoms of poisoning which would, in health, have followed the use of such large doses not being produced. Instead of such symptoms the patient experiences sensations similar to those which food produces under ordinary circumstances.

The toleration of large doses of alcohol has been discussed by Professor Beale,* in a recent communication to the British Medical Association; that gentleman considers that the action of this substance in acute disease is not exerted upon the nervous system to the same extent as would be the case in health; but is rather expended in checking the abnormal waste which is going on in the organs which happen to be especially affected. The excessive cell-growth, for instance, which is going on in a hepatized lung, may, he thinks, be arrested by this action of alcohol, and the exhaustive character of the disease thus changed to a more favorable type.

With all possible deference for the high authority on which this theory is propounded, it must be observed that it has grave defects. It proceeds upon the assumption that vital power is identical with rapidity of certain vital changes. Dr. Beale goes so far as to deny the name of "living matter" to all formed tissue, and to restrict it entirely to such portions of the organism as are in what he calls a "germinal" condition, that is, in one of rapid growth. Hence in his view the hasty and imperfect cell-growth, which is characteristic of inflammatory processes, is a typical instance of "vital action," which is here seen in excess, and requires to be checked.

Once more it must be repeated that there is no proof that excessive cell-growth is a true characteristic of vital power. The type of vital power for each organism, is the individuation

* On "Deficiency of Vital Power in Disease," and on "Support." *Brit. Med. Journal.* October 10, 1863.

proper to that organism: that individuation is a matter of balance and proportion: and excessive cell-growth indicates its destruction, more or less complete. It is a sign, not of a local excess of vital action or vital power, but of the escape of a certain part of the organism from the controlling conditions which constitute its true "life." It would be natural, then, rather to look for the essence of the morbid condition in the **default of some portion of the machinery** of co-ordination, than in the local extravagance of cell-growth: and to trace it to some defect in the blood or in the nervous system, and especially in the latter. We have already seen that damage to a nerve-trunk is probably a frequent cause of abnormal activity of cell-formation in the parts supplied by its branches. It would, therefore, as it **seems to** me, be more reasonable to suppose it possible that injury to the nervous system, of some kind which we do not understand, may be responsible for the occurrence of sudden and extensive inflammations of different organs, than to speak of a process in which a portion of the human frame is degraded to the condition of the lowest animals or vegetables as a dangerous *excess of vital action.* It would be rational to expect benefit rather from means which improve the condition **of** the nervous system, than from measures which would operate primarily on the local disorder; and to suspect that alcohol, when beneficial, acts in the former way, at least in great part. I cannot see the probability of Dr. Beale's suggestion that alcohol, as taken into the system in acute diseases, coagulates the albuminous matters of the blood, or renders them less fluid, and so hinders their permeating the tissues. The dilution would surely be far too considerable to admit of such an effect. And besides, **there is an** insurmountable obstacle to this explanation in the fact that precisely similar benefits (though less in degree) may often be obtained by the use of carbonate of ammonia, a substance which we certainly cannot suppose would act in the way referred to by Dr. **Beale.**

This general agreement of two such remedies as alcohol

and ammonia, in the effects they produce in acute disease attended with severe exhaustion, is very significant. Considering that their chemical relations to the blood are so different, it seems demonstrable that their common operation for good must be through their relations to the nervous tissues. And the same lesson is inculcated by the relief that each affords to more temporary conditions of depression, such as fainting, whether caused by hæmorrhage or by any other rapidly lowering agency. Here, also, as in febrile diseases, we find the intoxicating power of alcohol most notably diminished, so that delicate women, in the depression, for instance, of severe hæmorrhage after labor, have been known to take a pint or even more of raw spirit without evincing the slightest sign of intoxication.

But we are not left to negative evidence only for proof that both these stimulants exert a powerful influence on the nervous system. The removal of coma, the cessation of delirium, and the production of sound natural sleep, show plainly that the brain has been powerfully affected; and the cessation of the jactitations, or other convulsive movements that attended the state of depression, show equally that the nervous centers have been strongly influenced. In the chapters on Stimulation I tried to show that these are effects producible in many morbid conditions, as well by easily-digested food (such as soup) as by either alcohol or ammonia, though not so rapidly. And they are produced also, in other cases, by a variety of other stimulants—by turpentine, opium, sulphuric, nitric, and chloric æthers, various essential oils, &c. Of these, turpentine and sulphuric æther present the closest resemblance, in their action, to alcohol; and a very marked *toleration* of them is also set up in certain diseases.

For all these reasons, it appears to me that the proposal to class stimulants, at least provisionally, as a special variety of foods, is not devoid of rationality. In so much as they conserve the life of the organism, *not* by substituting abnormal for normal action in the principal organs, but by restoring

their natural functions, they seem to me to fill the *rôle* of
aliments as completely, in appropriate circumstances, as do
those substances to which the name of food is ordinarily given.
The opponents of this view would be perfectly justified in re-
jecting it if they could render any intelligible account of the
theory of stimulus. They have failed to do this. The doctrine of stimulus, as at present taught in class-books, is the
mere relic of an ancient doctrine which has no meaning now
for the teachers any more than for their audience, and should
be suppressed as a positive impediment to progress.

Surely there is something more philosophic in the view
that regards remedies of the stimulant class as *sustenance*
for the sick body, than in that which looks on them as missiles
directed against the imaginary entity, disease. That grand
old thought of the *vis medicatrix naturæ* would assume a
more elevated dignity, and would reflect more of the wisdom
of a Divine Providence, in the light of such an interpretation.
It is surely a worthier thought that for man in every varying condition of health there exists, somewhere, the appropriate *food*, than that under certain circumstances of misery,
produced by disease or starvation, human beings are allowed
to poison themselves with certain irritating substances which
they find ready to their hands, and which in some mysterious
way help them over their difficulties. The poor half-fed
Peruvian miner who knew that but for his regular daily
supply of coca he could not do his work, but that with it
he could live a long and laborious life in comparative comfort,
was not so extravagantly wrong when he called that substance
the "juice of life," and thanked his god for the good gift with
the same devout gratitude which the ancient Hebrew felt for
the "corn and wine and oil" of the promise.

It is strange, doubtless, that every, or nearly every stimulant we know, is in large doses a poison ; and a grave moral
might be drawn, were this the place for moralizing, from the
fact, that the evil always lies so near the good. But it is
still more strange to observe that to many who are unable

to believe in the possibility of alcohol, or sulphuric æther, acting in any dose as food, it never occurs that there is precisely the same difficulty in explaining the diametrically opposite actions of tea, of coffee, or of quinine, according as they are taken in **small or in excessive doses.** We are accustomed to think of these substances simply as tonic stimulants exercising a beneficial influence on nutrition; we are **apt to forget** that they **are** undoubted narcotic poisons when taken in excessive doses, and that excesses in tea and coffee **do actually** produce poisoning in a very considerable number of cases.

It has been shown conclusively, by Bernard and others, that substances which, both from experience and from their resembling closely certain important constituents of the **blood,** would be universally admitted to be food, when introduced in excessive quantities to the circulation conduct **themselves** exactly like poisons as regards their **stay in the organism.*** Thus, if an unnecessarily large quantity of **common salt** be administered to an animal at once, there will be no permanent increase in the quantity of that substance in the blood; the excess will not be fixed in that fluid, **but eliminated, just** like one of the metallic poisons. Thus, if **serum of blood** drawn from an animal be injected into the veins **of** another of the same species, **the albumen will not be retained,** it will be cast out of the **system in the urine. Thus, even** water cannot be permanently added in **any large excess to** the blood, but escapes **by some of the** various emunctories.

These instances are explained by M. Bernard on the ground **that chloride of sodium, albumen, and water respectively, are** constituents of the blood only under the form of certain **definite** "organic combinations," in ordinary circumstances, and that more than a certain quantity of them cannot, therefore, **be** retained. On the other hand, speaking of the occasional retention of the metallic poisons within the body (*e.g.* mercury, arsenic, antimony, &c.), he remarks that in

* Leçons sur les effets des substances toxiques, &c. **Par** Cl. Bernard. Paris, 1857.

fact these substances are *not* retained in the blood, but remain idly outside its current, as in the tissue of the liver, the bones, &c. These are substances which, according to M. Bernard, like the excess of true blood constituents, are "foreign to the organism:" they cannot, therefore, be retained in the blood. It is with considerable hesitation that one questions any argument of the illustrious French physiologist, but there is surely a fallacy here. True it is that where the metals in question have been given in poisonous doses over a considerable period, a part at least of the poison is apt to be extruded from the circulation and retained idly in the tissues of the liver, &c. But this does not account for another order of phenomena—those, namely, which are observed where small doses of arsenic, mercury, &c. are given during long periods as *tonics*, with the effect of manifestly improving the quality of the blood and tissues. Here it can hardly be maintained that the medicament acts solely as a "foreign" body, and contracts no "organic combination:" from day to day and from week to week we see the evidences that some powerful agent *is* operating upon the elements of the blood in a manner which indicates a very close, even if temporary, unison with them. Since the result **is** neither death nor impairment even, but, on the contrary, a restoration **of** the blood to the conditions of health, it is difficult to see upon what principle the title of ailment can be refused to a substance which produces these effects, and which appears **to play a** part (under the circumstances of the case) at least as important to nutrition as that of chloride of sodium in health.

And **if** even such substances as the undoubtedly poisonous metals are not "foreign," except in a relative sense, to the organism, still less are we justified in referring the class of stimulants generally to such a denomination. Many of them consist of, or contain, the very ingredients of which the blood is in part made up. With regard to the various diffusible hydro-carbons, in particular, it would be especially unjust, as it seems to me, to deny them definitely the title of foods upon the ground that they are foreign to the organism, unless some far more decisive proof were offered to us than we yet possess that such is the case. The proof of elimination "*en nature*," nay, even of elimination "*en totalité et en nature*," could not give certain demonstration of anything beyond the fact that the administration was untimely.

I may be allowed to say, also, that there is nothing in the present position of physiological chemistry which would justify the conclusion that our ideas of alimentation are finally and permanently settled. For instance, even so vitally important a question as that of the uses of the nitrogen of the atmosphere is by no means satisfactorily laid to rest. It is true that there is a large mass of evidence which appears to show that under the circumstances of ordinary health atmospheric nitrogen takes no active part in the vital processes: but it is far from being certain that this is the case in all pathological conditions; yet this is a fundamental point to establish, in deciding as to the possibility of maintaining the balance of material nutrition, in certain circumstances, without the aid of ordinary foods, or with an insufficient supply of them. For all we know, the administration of certain medicinal substances may effect important changes in the behavior of the organism toward the atmospheric nitrogen.

To recur to our former text, there is no existing definition of food, upon a merely chemical basis, which will satisfy us; and the only limitation of the word in which we are so far justified is that which confines it to substances which will avert death from inanition. It is no argument against this view that many of the substances which act as powerful stimulants can only maintain life, unaided, for a very limited time beyond that for which it might be supported by the mere combustion of the bodily tissues. The excess is sufficiently great, in many cases, to make all the difference between death from simple inanition and recovery. Under all these circumstances I submit that it is more rational to regard the stimulant class of remedies as exerting a food-action upon the organism, than to maintain a mythical doctrine of "stimulus and recoil" which has no basis, save in theories which are now exploded.

The practical effects of the whole line of argument which I now bring to an end may be shortly summed up in the following conclusions :

1. Narcosis, being a purely paralyzing process, does not fall within the number of remedial agencies, properly speaking.

2. Nevertheless, certain special forms of it may be employed for a strictly temporary purpose, as in the induction of anæsthesia by chloroform, &c.

3. The real therapeutic effect of the agents which we commonly call narcotics is a true stimulation: that is, when we employ them in proper medicinal doses.

4. When a particular symptom, e.g. pain, can only be relieved by narcotic doses of any drug, the medicine is probably altogether an improper one for the case.

5. In such instances the substitution of another stimulant, and especially the administration of appropriate food, will usually produce the desired effect, unless this be altogether unattainable.

6. It is erroneous to suppose that stimulation thus employed is followed by a depressive "recoil" in any other sense than feeding may be said to be so followed.

Such are the points which appear to me sufficiently made out for us to act upon : it will be seen that they involve a considerable change in the language, at least, which we now employ in speaking on therapeutical matters.

An immense task lies before the student of the action of medicines before the real operation of stimulants can be understood; and nothing short of the most patient and minute study of each member of the group, carefully separating the poisonous effects of large from the beneficial influence of small doses (and always keeping in mind that the words "large" and "small" are only to be read *relatively* to many other concomitant conditions of the experiment), can give us a full insight into the problem of stimulation. It is hoped that the suggestion of the close resemblance between the action of all true stimulants and that of foods may supply a clew which may somewhat assist further investigatio

It were earnestly to be desired that the mode of thought, so common among us, which deals with Life as if it were something capable of being measured by degrees more or less, might cease. I have endeavored, freely making use of the thoughts of others as well as of my own, to illustrate the difference between intensity of action and perfection of action. Life is not any special force, nor is it any "collocation of forces :"* it is not *in* the organism, for it is the very organism itself. Not the mere clay, indeed, however cunningly fashioned, but the thought of the Creator, binding together in wonderful relations the tissues of the material form and the forces of the surrounding universe. It is this which makes the individual—the Life; the mere fact of the aggregation of particular tissues, or of the development of particular forces, seems to me something widely distinct from it. Considering, therefore, the immense importance of the word "food," as indicating the part which we are intended voluntarily to contribute to the sustenance of life, it were to be regretted, I think, that physiologists should bind themselves irrevocably to a profession of scientific faith which seems to connect the idea of sustenance simply with the growth of structure,† and which would increase the tendency of students (never too weak) toward a partial and confined view of the phenomena of vital organization. And it would be still more to be regretted did we any longer shut our eyes to the facts which abound on every side, and which testify to an

* Bain, "On the Senses and the Intellect."

† *E.g.* Mr. Savory, in his excellent lectures on "Life and Death" (Smith, Elder & Co. 1863), p. 101.

occasional continuance of life under circumstances which do not square with our theories; contenting ourselves with the meaningless assertion that these effects were produced by the operation of a "stimulus," but that there *could* have been no real alimentation.

SPECIAL RESEARCHES.

INTRODUCTORY REMARKS.

UPON the title-page of this volume, the three substances whose properties we are now to examine are arranged in what may be called the order of their relation to the organism—Alcohol, which is the most important, being placed first, and Chloroform, which is the least important, last. This arrangement is based upon a rational principle: upon the fact, namely, that this is the order of the respective solubility of these three agents in the serum of the blood; and that it is the order also of their capacity for being retained in the blood and tissues.

It will be more convenient, however, for our present purpose, to study æther and chloroform before we turn our attention to alcohol: so that we may thoroughly understand what anæsthetic action is (of which these two substances give such marked examples) before we seek for evidences of a similar action, and attempt to trace its limits, in the operation of alcohol upon the organism.

17

SULPHURIC ÆTHER.

RESEARCHES ON THE PHYSIOLOGICAL ACTION OF SULPHURIC ÆTHER.

THIS inquiry consists of two parts:—I. An investigation of the narcotic effects of æther. II. An investigation of its effects when given so as not to produce narcosis.

I.—ÆTHER-NARCOSIS.

In describing the nature of æther-narcosis afresh, I shall abstain from describing such of my experiments as merely educe results with which the profession is already familiar; and confine myself to those which appear to establish new facts. Before commencing the relation of my own researches, however, it will be useful to sketch a *résumé* of the main conclusions established by the labors of others in the same field.

a. Local action of æther on the nerves. The researches of Serres, Longet, &c. have established the fact that the contact of liquid æther with a naked nerve rapidly deprives it of its functional capacity, apparently by producing a positive alteration in its intimate structure. It need scarcely be remarked that such an experiment as the above presents the æther to the nerve in a state of far greater concentration than that in which it reaches the subcutaneous nerves when applied to the unbroken skin.

b. Action on the general nervous system. This is only produced, to any noticeable extent, through impregnation of the blood with the narcotic, either in the liquid or the vaporous form. According to the law of toxic action, which M.

Bernard has so clearly expressed, effects upon the general nervous system can only be produced when the agent has penetrated to the finer ramifications of the arterial system. This stage having been reached, the nervous phenomena commence; and there is good reason to believe* that the nerves at the periphery are first paralyzed, and that the paralyzing influence gradually spreads to the centers. The general order in which anæsthesia of the surface is developed is universally allowed to be *from behind forward*, *i.e.* the nerves of the posterior extremities are first affected; and of these, the nerves of sensation somewhat earlier than the nerves of motion. The effects on motion and sensation proceed gradually to develop themselves in the more anterior portions of the body; the last part to lose its sensibility being, according to M. Bernard, the conjunctiva of the eye. It will be seen, hereafter, that my experiments reveal discrepancies with this account, in certain minor particulars.

The nervous centers begin to be affected soon after the earliest symptoms of peripheral sensory paralysis have appeared. According to M. Flourens† and Dr. Snow,‡ they are influenced as follows:—1. The cerebral hemispheres. 2. The cerebellum. 3. The spinal cord. 4. The medulla oblongata. No observer, hitherto, states the date of *commencement* of narcotic influence upon the *sympathetic* system, but all agree that this system survives the extinction of the functions of all the above-mentioned centers. According to the table thus arranged it will be perceived that the narcotized animal would lose—1. The local sensibility of extreme parts, and the control of certain muscles situated in those parts. 2. The intellectual powers. 3. The power of co-ordination of the locomotive organs generally. 4. The power of perceiving sensory impressions, even from parts little removed from the spinal centers. 5. The power of breathing. 6.

---

* Cl. Bernard, "Leçons sur les effets des substances toxiques," p. 330, &c.

† Comptes rendus de l'Académie des Sciences, tom. xxiv, pp. 251, 253, 310.

‡ On Chloroform and other Anæsthetics. London: Churchill. 1858.

The movements of vegetative life, e.g. of the heart, intestines, &c.

Concerning the correctness of this scheme, it may be said that among those most conversant with the operations of anæsthetics there has been little or no question, excepting what might be suggested by doubts as to the functions of the cerebellum, and the possibility of separating co-ordination of locomotive muscular action from the condition of the peripheral nerves.

c. The next statement of importance which has received general support from good authorities is, that circulation and respiration are similarly affected in the course of the induction of æther-narcosis; that their rapidity and force is at first augmented and afterward lessened. To this, and to the similar statement which has been made with regard to the action of chloroform, it will be seen that I have to take some important exceptions.

d. The effects on the development of animal heat have been studied with great care by MM. Dumeril and Demarquay,* and the conclusions arrived at by these observers— that the temperature is slightly raised during the very early, and depressed to the extent of 2.5 to 3 degrees (cent.) below the standard of health during the subsequent stages of prolonged narcosis—have, I believe, met with general acceptance. Upon this question I have made many observations, some of which appear to me to modify the general statement now quoted, but I shall refrain from dwelling upon them, feeling that more extended trials are necessary. Whatever the mechanism of the calorific changes in æther-narcosis may be, there is no doubt that they follow, for the most part, the changes in the activity of respiration and circulation. On the other hand, the latter changes will not, I believe, account for the whole of the effects on the production of heat; and these extra effects deserve careful investigation in the future.

---

* Recherches expérimentales sur les Modifications, &c. Arch. Gén. de Med. 1818, pp. 189, 332.

e. As regards the general order of the paralytic affections of the nervous system which have been enumerated, my observations present no very important novelty, except as regards the affections of the *sympathetic system*, but on the latter point I venture to believe that they indicate points of considerable importance.

f. With respect, also, to the distinction between the period of "excitement" and the period of "anæsthesia," it will be seen that my inquiries compel me to take an entirely different view from that commonly stated in books.

g. The question of the possibility of inducing anæsthesia, with full development of all its symptoms, by the introduction of æther into the body by other channels than that of the lungs, has particularly engaged my attention. The reader is doubtless aware that various observers, and especially M. Flourens and MM. Lallemand, Duroy, and Perrin, have stated positively that anæsthetics cannot be made fully operative except by offering their vapor to absorption by the lungs. My observations necessitate a very different conclusion.

1. *Experiments on the Human Subject.*

In the following experiments an apparatus was used which was a close imitation of that employed by Snow, and figured in his work on Anæsthetics. One or two ounces of æther were placed in the interior of the evaporation box, which was occupied by a spiral so arranged as to compel the atmospheric air to pass over a large surface of æther on its way to the mouth. The tube and face-piece were those of an ordinary Snow's inhaler; the latter was accurately adapted to the face.

Experiment I.—The apparatus having been charged as above described, a man aged forty, in sound health, and of muscular build, commenced inhalation, for the purpose of allowing an examination of his eyelids, as he had two days previously got a "spark" of something, from the furnace of a forge, into his eye, and the part was so unnaturally sensitive that he could not bear it to be handled.

No voluntary struggling took place, and the vapor did not appear to irritate the air-passages. Respiration, which at the commencement was 16 per minute, retained this rate during the whole of the first minute. The pulse (which at the commencement of inhalation was 74) mounted during that time to 96, and was very forcible in its beat. Sixty-five seconds from the first inspiration of æther the patient sat up and looked at me with a roguish leer for a moment or two. He then sank back and began to gabble incoherent nonsense with great fluency, and at first with perfect articulation: by the end of the second minute the pulse had risen to 104; respirations 18; eye somewhat congested, face of the natural tint, pupil apparently unaffected in size and quite sensitive to changes of light. There was now very perceptible diminution in the sensibility of the skin of the hands; there was also commencing rigidity of the muscles of the arms and fore-arms, and more decided stiffness of the legs. At the end of the third minute articulation had become confused; there was a copious flow of frothy saliva, which the patient made no effort to get rid of; consciousness was apparently lost, muscular rigidity was general and very strong, particularly in the muscles of the neck; face flushed and sweating, eyes very much congested, pupil contracted and insensitive. Pulse 98; respiration 28. At this moment an attempt was made to explore the injured eye, but the lids closed with spasmodic firmness at the first touch on the conjunctiva of their edges. Inhalation was continued for two minutes longer; at the end of this time muscular rigidity had disappeared, the patient was profoundly unconscious, the pupils dilated, and the con-junctiva perfectly insensitive; pulse 96; respiration 21, snoring. The eye was now explored, and the foreign body removed in less than a minute. The patient had completely regained consciousness at the end of seven minutes from the withdrawal of the æther inhaler; pulse 72; respiration 15. At this time, however, and for several minutes longer, there was still some feeling of numbness in the feet and in the

calves of the legs, slight dizziness, and a slight deficiency in
the co-ordination of the movements of the lower limbs in
walking. On examining the inhaler, three ounces of æther
were found to have been used.

Experiment II.—An ounce of æther having been placed in
the inhaler, the face-piece was made fast **to my own face by**
strips of adhesive plaster. My watch was placed before me
in such a position that I could easily see the movements of
the second-hand. With pencil in hand, I made a simple
mark upon paper for each fifteen seconds, so long **as** consciousness lasted. I had no assistant in this experiment.

With the exception of the odor of the æther being very
unpleasant, my sensations were highly agreeable, and no irritation of the air-passages was occasioned, although the outer
valve of the face-piece **was** left more than three-quarters
closed from the first. The first symptoms were those of
simple exhilaration, and warmth extending all over the body:
the pulse **was** somewhat increased in frequency and the heart's
action became **strong** and perceptible to myself. For more
than thirty **seconds, I experienced** no other feelings than
these. A **sense of** numbness and indistinct tingling then
began to affect the feet and spread upward with considerable
rapidity. Almost simultaneously, perspiration broke out on
the forehead, and I began to be dizzy, with a feeling as if the
room were spinning round. I felt a strong inclination **to**
laugh, and I believe I did so. It was now impossible for me
to see the movements of the second-hand of the watch, or
even the large figures; my limbs felt like lead, and almost
the last thing of which I was conscious was that my pencil
fell out of my hand, and that I could neither see **it** on the
floor nor move my foot to feel for it.

On recovering consciousness, I could not at first move any
of my limbs, and the room still seemed to spin round; the
face-piece was still firmly attached. It was some little time
before I could distinguish the figures on my watch; when I
had accomplished this, it appeared that thirty-five minutes

had elapsed since the commencement of inhalation. I was comfortably cool, but my face was damp with copious perspiration. There was still a sensation of numbness and tingling in all my limbs, and, on attempting to walk, I could not manage my legs. In less than five minutes more, I had perfectly recovered. It appeared that I had only made two marks upon the paper: this proved that I had become unequal to the requisite movements, or oblivious of the matter, before the forty-fifth second from the commencement of inhalation. All the æther in the apparatus had been used.

By comparing Experiments I. and II., which I believe offer a fair representation of the phenomena of æther-narcosis, we may arrive at a more clear idea of the order of events than is to be derived from the general descriptions of them given in such works as I have seen.

In the first place, it is evident that the ordinary phrase—"period of excitement"—as indicating an early stage, separate from the "period of anæsthesia," is likely to lead to serious misapprehension. The symptoms which would be considered as particularly indicating excitement were the muscular rigidity, the half-delirious movements, and the incoherent talk of the patient in Experiment I., and the disposition to hysteric laughter which I myself experienced. In both cases, however, it was noted that the symptoms referred to did not occur till the signs of advancing paralysis had already developed themselves: thus, in Experiment I., at the very moment when the patient sat up and began to resist somewhat, he took scarcely any notice of a sharp pinch, with which I tested the sensibility of his calf. His garrulous talk was quite incoherent: the muscular rigidity did not appear till an even later stage, when symptoms of paralysis were still more apparent. In my own case, symptoms of palsy of sensation appeared so early as half a minute from the commencement of inhalation, and distinctly preceded the emotional "excitement" which inclined me to laughter. The earlier mental state had been one of mere pleasant exhilaration of a tranquil kind, and perfectly different from the later condition, in which consciousness was evidently much impaired.

It is obvious, then, that it is incorrect to speak of the period of muscular rigidity, emotional disturbance, &c., as separate from, and prior to, the period of anæsthesia. On the contrary, they are clearly a part of the train of anæsthetic effects produced on human beings by æther (for, in this respect, all the cases that I have witnessed agreed perfectly with those above described). The undoubtedly voluntary struggles which sometimes precede the development of muscular rigidity are to be interpreted in the same way. They are the result of narcotic delirium, and are due to the paralyzing influence of æther on the cerebral hemispheres.

It appeared also, from Experiment I., that respiration and the pulse need not be affected in direct proportion to each other: thus, during the first minute of inhalation, respiration remained unaffected, while the pulse rose 22 beats; during the second minute, respiration only increased 2 beats, while the pulse rose to 30 above its frequency at commencement; during the third minute, the pulse only increased by 2 beats, while the respiration suddenly became increased by 10. After seven minutes of full anæsthesia, the pulse and breathing had both fallen slightly below their level at the commencement of inhalation.

Besides the above two cases, I have witnessed, or taken part in, the successful* inhalation of æther by human beings, thirty-four times: the majority of these inhalations were for experimental purposes merely. The following table represents, at a glance, the average effects upon the frequency of pulse and respiration at different periods of the anæsthetic process. The patients were all free from any discoverable disease of heart or lungs; 21 were males and 13 females; of the whole number 3 were under, and 31 over, the age of puberty.

* No cases are admitted to this list in which full anæsthesia was not induced.

	At commencement of inhalation.	At end of 1st minute.	At end of 2d minute.	At end of 3d minute.	At end of 4th minute.	At end of 5th minute.
Average frequency of pulse	74·5	92·7	109·8	110·2	94·3	69·3
Average frequency of respiration . .	23·0	23·0	24·7	26·3	18·9	15·67

On inspecting this table, it will be evident that though the general tendency of ætherization appears to be to quicken both respiration and circulation in the early, and lower them in the later stages of anæsthesia, yet the affection of these two functions does not by any means accurately coincide. And this, which is seen on an average of thirty-four cases, is more strikingly displayed by the evidence of Experiment I., which shows how far the variation from such coincidence may extend. The high rate of respiration which is marked in the table, as the average at the time of commencing inhalation, was owing to agitation and alarm on the part of some patients.

Of the 34 patients, 25 were affected with a symptom to which it is desirable to call attention, viz. flushing of the face (sometimes accompanied by sweating). This invariably appeared (when it occurred so as to be recognizable at all) in the very early stages of narcotism; in fact, it may be said to have been the first symptom of æther-narcosis. It was immediately followed, in every case, by an increased secretion of saliva; and it was obvious that both symptoms depended upon paralysis of the branches of the *cervical sympathetic.* This paralysis of the cervical sympathetic is one of the symptoms of alcoholic narcosis also, and in the case of that agent, as well as of æther, comes on very early.

2. *Experiments on Animals.*

A. Inhalation.

Experiment I.—A middle-sized, full-grown, and healthy rat was introduced quickly into a glass jar, of 1,260 cubic inches capacity, through which the vapor of 126 grains of æther had been thoroughly diffused (10 grains to the 100 cubic inches).

The animal instantly became very much excited, and ran ' round the jar, leaping up as if to escape. In less than one minute, it fell on its side; the respiration was perceived to be extremely hurried; the rat was, apparently, quite unconscious. Several clonic convulsions of the limbs now occurred, after which they appeared to become rigid. Five minutes after its introduction to the jar, the animal was only breathing by isolated gasps. It was now removed to the outer air, one or two gasping respirations ensued, and respiration then ceased. The thorax was opened, and the heart was observed to beat for three and a half minutes longer.

Experiment II.—An average-sized, full-grown, healthy white mouse was introduced into the same jar, charged with an æther vapor of the same strength as before.

Instantly, the animal gave one or two spasmodic leaps upward, and then fell on its side with rigid limbs and slight tremulous movements. In ninety seconds, it was removed from the jar, and almost simultaneously the breathing ceased. The chest was immediately slit open, and the heart seen to be motionless, except for a slight pulsation of the auricles. Irritation (by pricking) failed to set the ventricles in movement again.

Experiment III.—An active, full-grown rat, of medium size, was introduced into the same jar, which was charged with the vapor of 63 grains of æther (5 grains to the 100 cubic inches).

At the end of a minute the respiration was observed to be hurried and gasping; the animal moved restlessly, but was unable to co-ordinate its motions. One minute later it had fallen on its side, and was apparently unconscious; the limbs were extended and rigid: respirations 72. Four minutes from the commencement of the experiment respiration ceased; the thorax was now opened, without touching the abdomen, and a powerful æther odor was perceived; the heart was beating feebly, 60 per minute. The contractions were not quite stopped till twelve minutes later; the muscular tissue of the heart was then found to be but very slightly irritable, and irritation of the thoracic ganglia of the sympathetic produced no effect on the heart. The lower part of the intestines was filled with fæces; there had been no evacuation.

Experiment IV.—A healthy, active white mouse was introduced into the same jar, charged with a similar proportion of æther vapor.

In half a minute the animal fell on its side, and apparently became unconscious; the limbs were extended, and appeared rigid. Death by cessation of breathing took place in two minutes four seconds from the commencement of the experiment; the heart continued to beat for rather more than a minute, and then ceased, except the auricles, which continued to pulsate for a few seconds longer. The heart remained irritable much longer than in Experiment II.

Experiment V.—A healthy rat of average size was introduced into the same jar, which was charged with the vapor of 31·5 grains of æther (2½ grains to the 100 cubic inches).

The animal ran round the jar in a very excited manner, and tried to leap up the sides of it. In seventy seconds it began to stagger and fall about, as if drunk, and in less than two minutes it lay on its side. Being removed from the jar, however, it was found to be only partially unconscious: the hind quarters were quite insensitive, as *likewise the muzzle;* the conjunctiva, however, and the skin of the chest and fore limbs were quite sensitive. It perfectly recovered after five or six minutes exposure to the air.

Experiment VI.—The same rat was introduced into the

same jar, charged with the same dose as in the last experiment. The same symptoms occurred as before, but the animal was allowed to remain three minutes in the jar. On being removed to the air unconsciousness and anæsthesia were found to be complete. Left to itself, the rat remained in the same state for eleven minutes, and then began to recover: recovery was complete in about six minutes more. The anterior parts of the body, the fore legs and the conjunctiva, were the first to recover sensation; the hind quarters remained powerless and the muzzle insensitive for a considerable time after the animal had regained consciousness.

Experiment VII.—A full-grown healthy rat was introduced into the same jar, charged as before with the vapor of 2½ grains to the 100 cubic inches.

The same symptoms were developed as before, and the animal was allowed to remain in the jar fifteen minutes; at the end of this time respiration had entirely ceased. Quickly opening the thorax, I found the heart beating; it continued to pulsate for more than two minutes.

Experiment VIII.—A full-grown healthy rat was introduced into the same jar, charged with the vapor of 12·5 grains of æther (1 grain to the 100 cubic inches).

At the end of half an hour the animal (which had manifested some excitement at first, but afterward became quiet enough) was removed from the jar; it was fully conscious, but the hind limbs were powerless, and it could only feebly move the fore legs: the muzzle also was far less sensitive than it would normally be. These slight symptoms did not wholly disappear till more than twenty minutes later.

Experiment IX.—A healthy white mouse was introduced into a jar of 1,260 cubic inches content, which was charged with the vapor of 31·5 grains of æther (2·5 grains to the 100 cubic inches).

Death by cessation of respiration took place in nine minutes forty seconds from the animal's introduction to the jar: the heart continued to beat some minutes later.

Experiment X.—A healthy white mouse was introduced into the same jar, which was charged with the vapor of 12·5 grains of æther (1 grain to the 100 cubic inches).

At the end of twenty-five minutes the animal was taken out; it staggered and occasionally fell, but preserved entire consciousness, and in a few minutes had entirely recovered. When first taken out of the jar the hind limbs were weak, and somewhat less sensitive than the fore legs.

B.—Injections into the Peritoneal Cavity.

Experiment XI.—Ten minims of æther were injected into the peritoneal cavity of a full-grown rat at 11.42 A.M.

The animal was narrowly watched until 12.42, when, as not the slightest narcotic symptoms had appeared, 20 minims more æther were injected. In two minutes from this time the respiration rose to 120; four minutes later it was 150: there was still no unconsciousness or anæsthesia, no convulsive movements or muscular rigidity. By 1.20 the respiration had sunk to 100, and it kept at this level, or thereabouts, till 1.32, when the animal fell into a sleep, from which, however, it could be easily aroused, and in which it showed no symptoms of genuine anæsthesia. At 3.7, 30 minims more æther were injected. At 4.30 it was at length noticed that there was decided paralysis, both of sense and motion, in the hind quarters, and also that the muzzle was apparently somewhat insensitive; the conjunctiva was not in the least affected. The animal never got beyond this degree of ætherization. Next morning it appeared entirely free from any symptom of æther-narcosis, and ate food with a tolerable appetite in fact, appeared none the worse for the operation. Nor did any bad consequences follow.

Experiment XII.—A large and strong tom cat had 30 minims of æther injected into his peritoneal cavity at 3.15 P.M. The animal had been fed moderately at 8 A.M. At 3.50 the cat showed scarcely any symptom of narcosis, and one drachm more æther was injected.

18

3.52.   The animal, which had run away and crouched in a corner, was lifted to the floor, and found to be completely paralyzed in all its limbs; the face and the conjunctiva were still sensitive, but every other part of the surface was completely anæsthtized.   Respiration 36; heart's action flurried and rapid.

3.53.   The pupils are enormously dilated, but still sensitive to light; the cat still retains some consciousness; the paralysis of all the limbs is complete.   Respiration 36.

4.0.   The animal has now perfectly recovered its consciousness; it raises its head and stares round in a puzzled way, but has not the slightest power over its limbs, and is completely insensitive to pricking or pinching anywhere, except in the ano-gential region, the face, and the conjunctiva.

4.20.   Has recovered sensation and voluntary power in fore limbs; walks in a curious way, dragging the paralyzed hind limbs.   Respiration 32.   Heart's action 100.   After this experiment the animal recovered and no bad consequences followed.

A comparison between the two last described experiments leads to some interesting reflections.   From Experiment X. one might have been led to conclude the accuracy of the assertion of MM. Lallemand, Duroy, and Perrin, that full anæsthesia cannot be produced by the application of the anæsthetic agent to the body in the liquid form.   But in Experiment XI. we see an animal very many times larger than the subject of the former operation plunged into the profoundest degree of narcosis which is consistent with recovery, by the injection of a dose of æther only half as large again as that which the smaller animal received with so little effect.   It becomes an important question to decide "what became of the 60 minims of æther which were injected into the rat's peritoneal cavity?" It is impossible to suppose that it remained inert in the animal's belly.   The high degree of heat to which it would be there exposed would infallibly convert a consideralbe quantity of it into vapor, which must either be absorbed into the abdominal venous circulation, or distend the belly so enormously as to cause death by mere pressure on the viscera of the chest. Moreover, the *prolonged* contact of a considerable quantity of æther with the peritoneum could hardly fail to give rise to peritonitis, which clearly did not occur in this case. We can hardly doubt that the whole, or much the greater part, of the æther within a short time had entered the portal circulation; and we are obliged, as it seems to me, to explain the non-production of anæsthesia by supposing that the narcotic was in greatest part eliminated immediately from the lungs, in the form of vapor, before entering the arterial circulation, only a small portion reaching the nervous tissues. That elimination by the lungs might account for the phenomena observed is shown by the following experiment:

Experiment XII.—A large rough terrier, very active and in full health, had half an ounce of æther injected into his peritoneal cavity. The animal seemed frightened, but not particularly hurt, and when released ran about the room wagging his tail.

Three minutes after the injection the dog, which had been lying down for a few moments, got up and blundered across the room as if he had been drunk, tumbling down every now and then, and knocking himself against various objects. He now became highly excited, and falling on his side he began to snap at everything near him, and to knock his muzzle repeatedly on the stone floor. Placing him on his feet, I found that his hind quarters were quite paralyzed as to voluntary power and as to sensation; and, moreover, it was evident that the muzzle had lost all sensibility. The dog did not cry nor otherwise give the slightest sign that he was in any pain; it was obvious that the symptoms were those of partial æther-narcosis. An apparatus was now adjusted to the animal's face, by means of which the air expired from the lungs was made to pass through a solution of one part bichromate of potash in 300 parts of strong sulphuric acid; after three or four minutes had elapsed, the elimination of

æther was demonstrated by the formation of the emerald-
green oxide of chromium. From the decided effect produced
on the test-solution during the first few minutes it was obvious that the elimination was proceeding very rapidly; and
pari passu with this process the symptoms of anæsthesia
subsided. In half an hour from the commencement of the
test-process a very feeble reaction could alone be obtained,
and correspondently with this it was observed that only a
very slight degree of anæsthesia and muscular weakness of
the hind limbs remained. No' evil consequences whatever
followed this rough treatment of the peritoneum.

That the more rapid elimination of the æther from the
lungs, in the case of the dog and of the rat mentioned above,
did not altogether proceed from peculiarity in the organization of these animals, as compared with that of the cat, may
be suspected from the following experiment:

Experiment XIII.—A full-grown healthy rat had fifty
minims of æther injected into his peritoneal cavity at 12.12
P.M.

Half a minute later the respiration was found to be extremely hurried (104), and for a moment I supposed the
distention of the belly by æther-vapor might be causing compression of the diaphragm and lungs; but the flaccid condition
of the abdominal walls negatived this idea. The heart's
action could not be counted. At 12.13 the limbs were rigid
and tremulous. At 12.14 full anæsthesia was developed, and
the animal lay in a condition of incomplete coma; respiration
72, heart's action uncountable from its rapidity. From this
time the rapidity of the respiration sank fast, the animal
breathed by isolated gasps, and respiration ceased at 12.17
(five minutes after the injection). The abdominal walls were
still flaccid.

During the period of anæsthesia an apparatus was adjusted
to the rat's muzzle, by means of which the expired air was
carried through a similar test-solution to that employed in
the last experiment. Abundant evidence of the elimination

of æther was obtained. Immediately on the arrest of breathing the chest was slit open and the heart was observed pulsating feebly at the rate of 60 per minute. A very strong odor of æther was perceptible the instant the chest was opened. The heart continued to beat for 12 minutes, the auricles being the *ultimum moriens.* When these had ceased to act, no amount of irritation of the sympathetic ganglia in the thorax would re-excite contraction of the heart; but pricking the muscular tissue of the ventricle caused a feeble response for a short time. The effects of electricity were not tried. The intestines were observed to be dilated and unwrinkled; they lay perfectly still, and, although the lower bowel was full, no evacuation had taken place.

What was the circumstance which determined the non-production of anæsthesia in Experiment X. (which, by-the-way, had been performed on the very animal which was the subject of Experiment XIII.)? Looking at the results of Experiment XII. we must conclude that it was the ready elimination of the narcotic from the lungs (as already suggested) before it could reach the arterial circulation, which prevented its operation on the nerves and nerve-centers. But how are we to account for the difference between the results obtained in Experiments X. and XIII. respectively, which were made on the same animal, the total quantity of æther used being actually greater in the case where only slight anæsthetic symptoms were produced than in that in which rapidly fatal narcosis was induced? The only probable explanation would seem to be, that in the former case the æther being administered in divided doses of gradually-increasing size, but little of the first doses reached the nervous system, the greater part being eliminated from the lungs as fast as it entered the portal circulation; while in the latter the portal circulation was suddenly charged with an amount of æther which it was impossible for the lungs to eliminate so rapidly as to prevent a poisonous dose of the narcotic entering the systemic circulation, and thus operating upon the nervous system.

Experiment XIV.—A full-grown healthy rabbit had ℥i of æther injected into his peritoneal cavity; the animal had been moderately fed eight hours previously.

Full anæsthesia and unconsciousness were developed within three minutes; no muscular rigidity whatever was produced. This condition lasted for an hour and twenty minutes, and then gradually subsided; in two hours and a quarter from the time of injection the rabbit had perfectly recovered. During the period of most profound anæsthesia, the eyes were turned upward, and moved, with a convulsive movement, outward and inward alternately; the pupil widely dilated, the breathing slow (16), the heart's action very quick and thready. Not the slightest ill effects to the peritoneum seem to have been produced.

From the result of these and many other experiments which I have made, it would appear that absorption of the greater part, if not the whole, of a dose of æther injected into the peritoneum (and that within a short space of time) is the rule, and not the exception. It is to be noted that none of the animals mentioned above had been subjected to anything like a lengthened fast, and therefore none of them were in a condition specially favorable to rapid absorption; and it is probable that the dog, for instance, which was the subject of Experiment XII., if previously starved for forty-eight or seventy-two hours, would have received the narcotic so rapidly into his portal circulation as to have rendered it impossible for the lungs to carry on elimination so fast as to prevent the occurrence of the profounder degrees of anæsthesia. When all is said, however, there remains a difficulty in explaining the difference between the effects produced on the cat and on the dog, even making full allowance for the difference of their weights; more especially when we reflect on the fact, which will be elicited in the chapter on Chloroform, that the latter anæsthetic operates as effectually, when injected into the peritoneum, in the one animal as in the other.

It is obvious at any rate, that anæsthesia can be produced by the application of liquid æther to the absorbent surface of the peritoneum; and it is evident, also, that when this is the case the symptoms do not vary from those observed in æther-narcosis produced by inhalation. These are facts which, so far as I know, are novel; at any rate, they stand in opposition to the statements of most authors. And, what is equally satisfactory, it is proved by the results of these experiments and numerous others to which they have led me, that æther —provided, at least, that it do not remain too long in contact with the peritoneum—excites no inflammatory action in it. As this is the case, and as little or nothing in the way of shock appears to be caused by the mere first contact of the anæsthetic with the peritoneum, we have a very valuable mode of experimentation with æther ready to our hands.

We are therefore justified in taking an animal such as the cat, in which repeated experimentation has proved that æther injected into the peritoneum causes rapid anæsthesia, as the subject for comparative trials with different doses.

Experiment XV.—A healthy, full-grown, and very large cat (the subject of Experiment XI., from which it had perfectly recovered) had half a drachm of æther injected into the peritoneum. As had occurred in the former experiment, no traces, or only the slightest, of anæsthesia were produced by this small dose; but the application of the chromic acid test to the expired air gave indications, within one minute of the operation, that elimination by the lung was going on.

Experiment XVI.—Forty-five minims of æther were injected into the peritoneal cavity of the same cat (two days later) at 11.45 A.M. The animal took very little notice of the actual operation, though it resisted the preparations for it vigorously.

For the first five minutes nothing more was noticed than that the cat became somewhat dazed and drowsy: soon after this, however, it got up and began to run round the room, but its powers of steady locomotion were much interfered with, and it tumbled down repeatedly, and blundered against various objects. Eight minutes from the injection the hind quarters were completely paralyzed as to motion and sensation, and the muzzle was nearly insensitive. A few seconds later the fore limbs also appeared to have lost voluntary motion and sensation, and all the limbs were slightly rigid. Beyond this point, narcosis did not proceed: the animal never lost consciousness, or the power of moving its head, or the sensibility of the conjunctiva; the pupils were never fully dilated, and they remained, throughout, sensitive to light. Recovery commenced about ten minutes from the time of injection, and was complete by 12.2, so far as regarded the anæsthetic symptoms.

Experiment XVII.—The same cat, being in full health, and having eaten food with relish seven or eight hours previously, had ʒi of æther injected into the peritoneal cavity, at twelve o'clock (the day after the last experiment).

12.1. The animal ran quickly round the room, and seemed much excited.

12.2.30. Movements uncertain; the cat falls down repeatedly; respiration rapid and panting, circulation too rapid to be counted; pupils contracted; hind limbs partly paralyzed.

12.4. Complete paralysis of sensation and motion and partial unconsciousness; eyeballs turned up, and moving, with a rhythmic alternation, outward and inward; pupils dilated widely, but still sensitive in some degree to light; conjunctiva insensitive; respiration 36; circulation extremely rapid. This was the point of deepest narcosis, at which the animal remained stationary for nearly ten minutes.

12.14. Respiration 32. Circulation 140. The cat raised its head and looked about; the conjunctiva was fully sensitive; muzzle insensitive; complete paralysis of sensation and motion in all the limbs.

Recovery was complete at 12.31.

The cat, which was the subject of these last experiments, was not operated on any further with æther.

Experiment XVIII.—A large, full-grown cat, near about the same size as that one used in the other experiments, had ℥iv of æther injected into its peritoneal cavity.

Respiration *immediately* became very hurried; circulation 160. The animal collapsed, so to speak, and fell on its side senseless, with the pupils widely dilated, the eyeballs convulsed; the rectum and bladder were involuntarily evacuated. In less than eighty seconds respiration had ceased, and on **immediately opening the chest**, the heart was found beating very languidly and irregularly 45 per minute. The contractions ceased in less than a minute, the auricles moving a little longer **than** the ventricles. The smell of æther was strongly perceptible in the cat's breath within forty-five seconds **from the** moment of injection, and on opening the thorax a powerful æther-odor rushed out. The heart's irrita-**bility was evidently very** much lessened by the poison, and **very** soon departed altogether.

On analyzing the above-narrated experiments we may come, I think, to some conclusions which it will be as well at once to state.

1. In the production of æther-narcosis important differences may be noted in the order of the symptoms according to the rapidity with which the blood receives the higher degrees of saturation. The general principle asserted by Snow is vindicated by experimental fact.

2. In æther-narcosis induced by the inhalation of an atmosphere *weakly* impregnated with the vapor, the narcotic effects consist of a paralysis which spreads from periphery to center, which involves the brain, the sensory, the motor, and the sympathetic system to nearly an equal extent; the sympathetic phenomena probably appearing slightly the earliest, and the sensory affection slightly preceding the motor.

3. **The same results are produced by the injection of a** moderate dose **of liquid æther into the peritoneal** cavity **or** into **the interior of the digestive canal,** *unless* the anæsthetic

should chance to be eliminated by the lung so rapidly as not to reach the arterial system in any considerable quantity— an occurrence which sometimes takes place.

4. In either case, if the process do not extend over too long a period, it tends naturally to recovery. The too great prolongation, however, even of this, which may be called the typical form of æther-narcosis, tends to produce death, by paralyzing the respiratory movements through its effects on the medulla oblongata.

5. Under the circumstances of *very rapid* saturation of the blood with a large dose of sulphuric æther, the course of narcosis is materially disturbed, and tends to the immediate production of dangerous or even fatal symptoms, which differ from those observed when an animal gradually sinks into death by apnœa as the result of the protracted operation of less overwhelming doses.

6. The statement made by Dr. Snow,* and repeated by the Committee of the Boston Medical Improvement Society,† that æther is altogether incapable of causing sudden death by paralysis of the heart, is considerably invalidated by the result of several of the experiments with strong atmospheres, supported as they are, moreover, by the analogous effects observed to result from the injection of very large doses into the peritoneal cavity. (*E.g.* Experiment XVIII.)

7. The statement that circulation and respiration are affected in direct proportion to each other, and that both these functions are rendered more active in the earlier, and depressed in the later, stages of ætherization, is too vague, and is, moreover, inaccurate ; for it not unfrequently happens that the circulation is greatly quickened, while the respiration remains almost at its normal frequency, or, at any rate, is very much more slightly hastened. And the idea that so long as the action of the heart has not been diminished in frequency

---

* Op. cit.

† Report of a Committee of the Boston Society for Medical Improvement on the alleged dangers which accompany the Inhalation of Sulphuric Ether. Boston, 1861.

that organ may be held to have escaped or resisted the narcotic action of the æther* would appear to be a serious mistake. Extreme quickness of the pulse is very well known to
be a symptom of great cardiac debility; it is considered to
afford this indication in the various "adynamic" fevers. And
by analogy we find this phenomenon very frequently indeed,
almost constantly, developed as a symptom of very acute
anæsthetic poisoning, which is yet not quite severe enough
to arrest the co-ordinated movements of the heart altogether.
In many of the experiments now detailed, and in many others
which I have made, it has been noted that great rapidity of
circulation was the not uncertain harbinger of a rapid and
shock-like fall of the pulse-rate; but in all these cases it was
obvious that the rapidity as well as the subsequent slowness
were the direct consequences of a paralysis of those portions
of the nervous system which regulate the heart's action. It
is therefore erroneous to speak of the phenomena of cardiac
"excitement" as belonging to the "pre-anæsthetic" stage.

8. If this be the case, it would seem to follow that excessive
rapidity of circulation in æther-narcosis must be attributed
to partial paralysis of the sympathetic system, for the following reasons: We have evidence that the sympathetic system
is affected, even in gradual anæsthesia, by the paralyzing influence of æther at an earlier stage of narcosis than that in
which the medulla oblongata is considerably influenced. It
can hardly be doubted that the increased effects of a more
rapid impregnation of the blood with the narcotic would tell
more rapidly on the sympathetic nerves than on the medulla
oblongata, or on the pneumogastric branches, on account of
the peculiarly intimate connection of the former with the
arterial tree. And any one who has experimented on animals
who have died from apnœa must have been struck with the
powerfully exciting influence which can be brought to bear
on the heart by rough handling of the thoracic sympathetic

* Du rôle de l'Alcool et des Anæsthésiques. Lallemand, Duroy, and Perrin. Paris,
1860. P. 259.

ganglia, such as must inflict injury on them not inferior to that which a paralyzing narcotic might cause. Under these circumstances, it seems desirable to study carefully the various evidences of sympathetic paralysis, which present themselves in the course of æther-narcosis.

(a) The earliest symptoms of sympathetic paralysis are to be traced in the flushing of the face, which, according to my observation, is very frequently, if not always, to be noted in the human subject of æther-narcosis, and which is usually attended by a perceptible perspiration; this symptom is also noted in the early stages of alcohol-narcosis.

(b) Simultaneously with, or soon after the appearance of these symptoms, there is usually a more or less copious secretion of saliva. In several human patients, and in many mammalian animals, I have observed this phenomenon; and this not merely when the anæsthetic was inhaled, but also when it was injected into the peritoneal cavity, the rectum, or the stomach.

(c) The influence of sympathetic paralysis upon the *circulation* varies extremely, according to the rapidity with which the blood becomes impregnated with æther. When this agent enters the circulation slowly, and narcosis is gradually produced, the elevation of the pulse-rate is at first comparatively small, and *may* never reach a higher point, unless the poisoning be carried to a fatal issue. On the other hand, when a large dose is rapidly introduced into the blood (*e.g.* Experiment XVIII.), the circulation soon becomes extremely hurried, at the same time that the force of the heart's movements is greatly impaired, and this excessive hurry of circulation not unfrequently continues up to a moment when it is abruptly exchanged for intermittence or even for complete arrest. I shall quote here the testimony of MM. Lallemand, Duroy, and Perrin (in describing one of their experiments*) as a confirmation, certainly by unprejudiced witnesses, of the state-

* Du rôle de l'Alcool, &c., p. 368.

ment that extreme rapidity of pulse coincides with a state of
great depression. These gentlemen administered æther, by
inhalation, to a dog, in divided doses. After ten minutes inhalation, partial insensibility having been produced, *and a
profuse flow of saliva taking place* (one of the symptoms,
already mentioned, of sympathetic paralysis), the apparatus
was charged with a fresh dose, and two or three minutes later
profound anæsthesia was produced: at this moment the circulation was 184; only a very slight diminution of this high
rate had taken place twelve minutes later, when the respiration had already greatly diminished in frequency and the
force of the heart's beat was much impaired. Even when
respiration ceased, the beating of the heart was 130 per
minute. Two minutes later it ceased.

The *highest* degree of paralyzing action upon the sympathetic nerves which are distributed to the heart is that which
causes instantaneous arrest of the co-ordinated movements of
the heart without any preliminary quickening. This effect
is seldom if ever witnessed in æther-narcosis, though it has
repeatedly been produced by chloroform, as will be hereafter
mentioned. Ordinarily, with æther, the effect of a very large
dose (injected, for instance, into peritoneum or pleura) is to
set the heart *running* rapidly for a few moments before it
stops.

(*d*) An abnormal formation of sugar by the liver, leading
to artificial diabetes (the excess of saccharine matter being
eliminated by the kidneys), has been noted by M. Bernard[*]
and by Dr. Harley;[†] it would appear from their researches
that this symptom is far more readily produced when æther
is presented directly to the portal vein or its tributaries for
absorption, than when it enters the circulation by way of the
lungs, as in inhalation. This phenomenon has received different explanations, according as the disturbance of the normal

[*] Leçons sur les Effets des Substances Toxiques, pp. 413–435

[†] On the Physiology of Saccharine Urine. Brit. and For. Med.-Chir. Review.
July, 1857.

action of the liver has been regarded as depending on an increased nervous influence or evolution of proper gland-tissue.

In the view of M. Bernard this effect upon the liver function is a thing altogether apart from the *anœsthetic* operation of æther upon the organism. It is due to a direct excitation, compelling a more rapid evolution, of the proper gland-tissue of the liver. He affirms that æther is *in all doses*, a direct excitant of the functions of the abdominal viscera by its local action; and he contrasts its behavior in this respect with that of alcohol, which, when applied in a concentrated form to the surface of the alimentary canal, *arrests* secretion. Between these two substances, which as to their anæsthetic effects are very similar, there is a radical distinction and even opposition as to their local operation on the gland-tissues; a circumstance which seems to separate the latter from the number of the *narcotic* phenomena produced by either alcohol or æther.

It must be remarked, however, that the only proof of this fundamental opposition given by M. Bernard, is the observation that *pure* alcohol and *pure* æther respectively produce, the first a diminished, the second a very copious, flow of the gastric juice, when taken into the stomach; and that the first causes an alteration in fragments of gland-tissue which have been steeped in it, which the latter does not cause in similar portions submitted in the same way to *its* influence. This argument appears to be somewhat unfair; since we are not called upon to consider such a phenomenon as the direct application of pure alcohol to the gland-tissue of the liver, an occurrence which could never take place in the living body. And on the other hand, it has been shown by Dr. Harley* that alcohol, diluted, but still in anæsthetic doses, when injected into the portal circulation *does* exaggerate the glycogenic action of the liver.

If now we consider the fact that æther, injected into the portal vein, or reaching it by rapid absorption from the stomach or intestines or peritoneal cavity, must, during the

---

* Loc. cit.

whole of its stay in the portal circulation, be in close proximity to the sympathetic branches which surround the ramifications of that system—we can hardly doubt that this volatile and penetrating substance must powerfully affect those nerves; and believing this it is scarcely possible to imagine its influence to be other than a paralyzing one, since a very small amount of æther or of its vapor is enough to produce such effects when *locally* applied to nerves. The same effects, as I cannot but suppose, would be produced, though in a less marked degree, upon the pneumogastric nerve-filaments in the liver: at least this appears more probable than the ingenious suggestion of Dr. Harley, that these nerves are *stimulated*, and by conveying this stimulant impression to the medulla cause a reflex excitement of the liver through the splanchnic nerves. Moreover, the investigations of Dr. Pavy,* if they have not finally disproved the possession of a true vital sugar-forming function by the liver, seem to point in the strongest way to the conclusion, that the appearance of sugar in the general circulation to any considerable extent implies an abrogation rather than a stimulation of vital powers in that organ, and with much force, also, to the conclusion that it is the extension to the sympathetic nerves of this devitalizing process which is answerable for the occurrence of diabetes.

For these reasons I am unable to accept even the high authority of M. Bernard for the statement that diabetes produced by æther is no part of the anæsthetic effects of that agent. On the contrary, I believe it to be a very characteristic part of the train of narcotic phenomena. The production of diabetes, however, does not immediately follow the action of æther on the hepatic nerves to which I believe it owes its origin. In each of the following experiments a quantity of æther was employed which assuredly would have produced very decided diabetes had life lasted longer; it will be seen, however, that from two to four hours was insufficient to bring about the result.

* On the Nature and Treatment of Diabetes. London: Churchill. 1862.

Experiment XIX.—A full-grown and healthy rabbit, which had been fed four hours previously, had ℥ss of æther injected into its peritoneal cavity, at 11.50 A.M. At 11.55 the animal was much excited and decidedly inebriated, for though active enough in running, its movements were uncertain and ill co-ordinated: moreover, there was an evident diminution in the **sensibility of the hind** limbs. **It remained** in about the same condition, except that a slight degree of motor palsy was de-veloped in the hind limbs, at 12 o'clock. An apparatus was now adjusted by which the breath was conveyed through a test solution of chromic acid. At 12.15 the solution first began to experience a slight change of color; this rapidly deepened, and in less than two minutes became changed to a brilliant emerald green. At this moment the condition of the animal was that of partial but decided narcosis: there was considerable motor and almost total sensory paralysis of the hind limbs; the rabbit was apathetic, and, though per-fectly conscious, appeared to be quite *deaf;* for whereas it was originally an especially timid creature, it paid not the slightest attention to the loudest noise made close to its ears. The circulation was 300 per minute, the respiration 47. One hour later the animal remained as nearly as possible in the same condition in every respect; the breath was again passed through a chromic acid solution, and almost at once produced **a** notable change of color, which rapidly progressed till the characteristic emerald green was developed. At this time the animal's belly was carefully squeezed, so that the contents **of the** bladder were expelled as completely as possible; the urine gave not the faintest traces of reaction with Trommer's test, even on prolonged boiling. At 2 P.M. the elimination of æther by the lungs was no longer demonstrable even by ten minutes' application of the chromic acid test, but a faint odor of æther still lingered in the breath. Respiration 38; circulation 204. No outward sign of anæsthesia now re-**mained, save a little** drowsiness, the sensibility and motor **power appeared** perfect. The animal was now killed instantaneously by opening the thorax and dividing the great vessels; the **peritoneal cavity** contained a few drops of clear serum having a slight odor of æther; there was a very **moderate** degree of injection **of** the subperitoneal vessels of the intestines. The bladder was full of urine; this was removed **and tested for sugar, but not** the least trace of that substance could be found.

Experiment **XX.**—A full-grown but **rather small** and weak rabbit had ʒi of **æther injected into his** peritoneal cavity at **1.33 P.M.** In rather less than one minute the fullest unconsciousness and anæsthesia were developed; the odor of æther was also plainly perceptible in the breath. Respiration 64, circulation 280. At 1.35 the respiration was 87, **circulation 300, pupils dilated and** insensitive **to** light. **The animal** remained in a state of deep narcosis, the depression of the nervous system appeared slowly to increase up to 3.35, when involuntary evacuation of urine took place; a little of this **was** collected and tested, but no trace of sugar was found. Respiration was now 78, circulation 200, very weak. Soon after this signs of recovery began to be perceived, and at 4.15 the animal had perfectly regained consciousness, and the sensibility and voluntary power of the fore limbs. **Circulation** 180, of good strength; respiration 56. ʒij of æther were now injected into the right pleural cavity, the breathing almost instantly became hurried, then gasping, and in less than two minutes had ceased. On quickly opening the chest the heart was found passive, excepting a little fluttering of the auricles: nor could it be excited to contraction by irritation of the thoracic ganglia; pricking the wall of the ventricle, however, re-excited contraction for nearly fifteen minutes. During the short interval between the injection and the cessation of breathing the smell of æther was strongly developed in the breath. The bladder contained urine **which** must have been secreted within the last forty minutes of life; this was tested, but no sugar was found.

Experiment **XXI.**—A full-grown, healthy, and remark-

19

ably strong rabbit had ℥iss of æther injected into his peritoneal cavity at 1.47 P.M.

At 1.49 no symptoms of anæsthesia had appeared, but the
belly was enormously distended, obviously by the vapor of
æther, and the breathing was embarrassed by the compression of the contents of the chest.

At 1.51 the distention had considerably subsided, and
symptoms of anæsthesia were apparent; the hind quarters
were paralyzed as to sense and motion. At 1.54 the paralysis of sense and motion had extended to the anterior limbs;
the animal was still perfectly conscious, and moved the head
and neck with freedom. Circulation 268, respiration 72.
Anæsthesia never advanced beyond this period; at 4.4 the
animal had almost entirely recovered.

℥i of *chloroform* was now injected, *per rectum;* in less
than two minutes respiration had risen to 120, circulation to
314; there was great excitement, and loss of co-ordinative
power over the limbs. At 4.9 there was complete paralysis
of sense and motion in the hind limbs, but the excitement
continued, and the animal dragged itself rapidly about the
room. This condition lasted for more than an hour, the excitement then passed off, but the animal remained stupid and
with its hind limbs paralyzed. At 5.48 the rabbit, having so
far remained to all appearance in the same condition as has
been described, was suddenly attacked with a violent parox-
ysm of tetanic convulsions in which it died : rigor mortis
occurred, apparently without the slightest interval, after
death, fixing the body in the attitude of extreme opisthotonos.

The bladder of this animal had been completely emptied
just previously to the injection of chloroform (one hour and
three-quarters after the injection of æther); no sugar was
found in the urine. On opening the body after death, the
bladder was found distended with urine, which contained a
little blood; this was tested for sugar, but none was found.

Experiment XXII.—A full-grown, healthy rabbit had ℥ij
of æther injected into his rectum at 3.5 P.M.

At 3.7 the animal was extremely drunk, and his hind limbs were partly paralyzed, but he moved about actively enough. At 3.20 the rabbit was still much in the same condition, but rather more stupid and drowsy. Respiration 92, circulation too rapid to be counted. ʒss more æther was now injected into the rectum.

3.21. Profound anæsthesia: the rabbit lies comatose, with widely dilated pupils; respiration 32, gasping; circulation *too quick to be counted*.

3.22. Respiration has stopped. The chest is instantly slit open; a faint fluttering of the auricles is the only remnant of cardiac action. Irritation of the sympathetic ganglia in the thorax produces no increase of contractions. Pricking the muscular structure of the ventricle only gives rise to local and partial contractions.

Examination of the urine contained in the bladder (which must have been secreted within the last three or four minutes of life, as the bladder had been thoroughly emptied by pressure just previously to this and was now nearly full) showed no traces of sugar.

On a superficial examination of the cases above described it might at first seem that they indicated that paralysis of the sympathetic was *not* the cause of the artificial diabetes which has been observed by Bernard and Harley to follow the introduction of æther into the portal circulation. A high degree of palsy of the sympathetic system generally was probably produced in all these four experiments, yet no diabetes occurred during the very considerable periods of time for which the animal survived the administration of the narcotic. It must be remembered, however, that upon the theory that diabetes results from a withdrawal of the nervous influence which would ordinarily prevent the liver from forming a sufficient quantity of sugar to overcharge the circulation and cause the elimination of that substance by the kidneys, the process would naturally be a gradual one; and it would be some time before the blood would contain such a quantity of

saccharine matter as would necessitate elimination. On the
other hand, if we could suppose with Bernard that the tissues
of the liver, or with Dr. Harley that the pneumogastric
nerve-filaments, are directly *stimulated* by the local application of æther to them through the portal circulation, it seems
to me that the process ought to be far more rapid, and the
production of diabetes to occur much within the periods of
time which intervened, in the above experiments, between
the administration of the narcotic and the occurrence of
death. There can be no pretext for supposing that **the doses**
employed were insufficient for the purpose, for the rabbit is
an animal particularly easily affected in this way, and any
one may readily convince himself experimentally (as I have
done) that a much less quantity of æther will produce diabetes
within a few hours, if life be prolonged.

So far, then, from thinking that the production of diabetes
by the local action of æther upon the liver is a phenomenon
removed from the category of narcotic effects, I look upon it
as affording an interesting and valuable extension of our
knowledge of these effects. The persistence of a saccharine
condition of the urine for very many hours, frequently for a
day or two, after the apparent cessation of anæsthetic symptoms, which has been frequently observed, is, in my opinion,
inconsistent with the notion of a true stimulation of the liver;
it is rather to be ranked, as it appears to me, with the **persistent diarrhœa which is observed in recovery from** poisoning with doses of the Calabar bean sufficient to produce symptoms of **impending** cardiac palsy, and which I strongly suspect
is due **to paralysis of** the intestinal branches of the sympathetic
nerve.

(*e*) Finally we have to speak of the changes in the pupil
which are produced in æther-narcosis. My observations on
this point, as far as regards the human subject, would be
decisive were it not for their necessarily limited number. In
every case of inhalation which I have witnessed in man, the
first effect of æther has been to *contract* the pupils : and if a

high degree of narcosis was produced, with snoring respiration, &c., *dilatation* supervened. With regard to animals, I am not able to speak positively; but this I may safely say, that in every fatal administration the pupil has become widely dilated at the approach of death. (The reader is referred to Note B, for the statement of my opinion as to the causes of the sequence of pupil-changes in **narcosis**.) **The** affection of the pupils is one of the earliest narcotic symptoms to depart, in recovery from ætherization.

We have now gained an intelligible view of the phenomena of æther-narcosis, and I trust the reader is convinced of the necessity for discarding the idea that this process consists of two stages, one of excitement and the other of anæsthesia. The whole process is essentially one; it consists, both in its early and its later stages, of paralysis of the various sections of the nervous system. There is, however, an action of æther upon the organism which is quite different from this. It is produced by doses altogether insufficient to produce paralysis of any part of the nervous system, or rather, to speak more exactly, by the presence in the blood of a proportion of æther which is not sufficient to convert the circulating fluid into a paralyzing agent. I shall now adduce evidence of the reality of this kind of physiological action.

II.—THE STIMULANT ACTION OF ÆTHER.

A.—On Man.

In order to ascertain **whether a** perfectly non-narcotic action of Sulphuric æther was obtainable, I made the following experiments on myself:

Experiment XXIII.—At 10.35 P.M., four hours after taking food, having brought myself into a **state** of quiescence by sitting still for some time, I swallowed one drachm of æther (Sp. gr. 750) suspended in four ounces of a thickish mucilaginous mixture. It was necessary to drink it gradually, as the taste is so disagreeable to me as to be highly provocative of vomiting, but in about three minutes the task was accomplished. A very slight feeling of warmth at the epigastrium was the only symptom noticeable during the next five minutes. At 10.38 thirty seconds, my finger on my pulse (which had been 74 and rather weak at the commencement of the experiment, and had not yet varied,) was sensible of a decided increase in the fulness of the beat, and during the next sixty seconds the number of pulsations was eighty. Simultaneosly with this improvement of the pulse a sense of lightness and comfort, and the removal of a feeling of weariness which I had previously had, was very obvious. During the next minute the pulse rose to 84, and its firmness and fullness somewhat increased. During the next ten minutes the pulse was as follows: 86, 88, 88, 88, 86, 88, 86, 85, 85, 84. The feeling of bodily comfort continued unabated until I went to bed at 12.30 P.M.; the pulse at that time being 79: at no time was there any flushing of the cheeks or uncomfortable heat of the face, or any local numbness or tingling. Both on going to bed, and also the next morning at 9 A.M., and at 6 P.M. of the same day, tests were applied to ascertain if any elimination of sugar by the kidneys was taking place, but with a negative result. I pass over successive experiments with increasing doses (made at three or four days' interval from each other), the results of which were similar to those of Experiment XXIII. in order to come to—

Experiment **XXIV.**—At 10.35 P.M. four hours after taking dinner, I swallowed ℥ijss of æther (Sp. gr. 750) suspended in six ounces of a mucilaginous mixture; the drinking of this very nauseous dose occupying about ten minutes. It was scarcely finished before I experienced a considerable and rather uncomfortable heat at the pit of the stomach, and the pulse rose rapidly in frequency, and, at first, in force. At 10.50 the pulse was *bounding*, rather than full (104), there was decided giddiness, great heat and flushing of the face, and perspiration on the brow; the lips also felt stiff, and there was a pricking sensation in them; the sense of the heart's action was painful, as in an attack of palpitation. The mental condition was one of slight confusion. At 12.30 this feeling of giddiness and confusion had hardly disappeared, pulse 96. At 8 A.M. on the following morning I awoke with a headache and nausea, which soon passed off. At 6 P.M. I obtained decided though slight traces of the elimination of sugar by the kidneys, by the use of the tartrate of copper test.

N.B.—In all the above experiments the use of alcohol was avoided on the day of the experiment, and on the day following.

From the above experiments (which I repeated with substantially the same results) I considered myself justified in concluding that a dose of æther about 3i might be safely considered non-narcotic, and that 3ijss was decidedly narcotic, to me at least. It was only by the use of the latter dose that I had experienced any of the symptoms such as flushing of the face and *palpitation* of the heart, confusion of intellect, and especially the production of artificial diabetes. There was nothing in any of the symptoms produced by the first dose which could be considered abnormal; in the last experiment there was much that was distinctly so.

Experiment XXV.—A. B. æt. 42, stone-mason; a sufferer from excessive dyspnea, which comes on periodically, and is due to laryngeal spasm, the result of pressure exercised by an aneurism on the recurrent nerve. This man is in the habit of taking, on the average, two drachms of æther a day, not in the form of Hoffman's liquor, but merely suspended in mucilage. The effect of a single dose of this kind is, in most cases, instantly to relieve his spasms (which come on every evening). It occurred to me that it would be interesting to examine the urine for sugar, and this was done daily, and at various times, for some time; but not the slightest trace of sugar was ever detected, except on one occasion, viz.: after

a particularly severe and obstinate fit of dyspnea which had
greatly prostrated him, but in which he had not taken a
larger dose of æther than usual. This man was always extremely pale and depressed, and his pulse was feeble, for some
minutes previously to his seizure; on the actual occurrence
of the spasm the pallor of countenance gradually became
mixed with a livid hue. The effect of the æther was to
rapidly restore his pulse to its normal strength, but not to
hurry its action; and simultaneously with this spasm always
subsided. If the pulse were *not* restored, and the livid pallor
of face persisted, as occasionally happened, the spasm remained unrelieved. There never was any flushing of the face
or confusion of intellect produced by the medicine. Latterly
I have lost sight of this patient ; but he was more than three
months under my care. The same dose of æther regularly
produced the same effect, during the whole of this period.

Experiment XXVI.—A woman, æt. 44, the victim of
chronic alcoholism, applied for relief from constant muscular
tremors and from frequent attacks of giddiness, to which she
was subject. For three weeks she was made to take 3iss
of æther per diem in three doses. The urine was repeatedly
analyzed, and no sugar was found; the woman never experienced any flushing of the cheeks and giddiness, &c. A very
considerable abatement of the symptoms was experienced;
and this was permanent, although the case was beyond the
hope of a complete cure. The rapid effect of each dose of
æther, in steadying the tremulous muscles and rendering co-
ordination of movement (with the hands especially) more efficient, was very striking. There was no sort of after-depression following the first happy effect of the stimulant dose, but
merely a gentle subsidence of this effect. I ought to state
that this patient had entirely given up hard drinking, and
confined herself to a moderate quantity of beer daily, some
time before she came under my charge.

Experiment XXVII.—J. B. a warehouse porter, æt. 47
who had lived a very fast life, and drunk excessively hard

(of spirits principally), applied to me at Westminster Hospital, suffering from entire sleeplessness, great anxiety of mind, and hallucinations. The night before he came to see me he had of his own motion taken sixty drops of laudanum, with the effect of increasing rather than diminishing his restlessness and misery, and leaving him very sick and nauseated some hours afterward. For some weeks past he had taken only two pints of beer a day. He was ordered to take ℈i of Sulphuric æther (in mucilage) at night, and for the first time for ten days he obtained a continuous sleep of five or six hours, and awoke refreshed and comparatively free from the tremulousness and horrors. Subsequently this patient was recommended to take food and a glass of beer immediately before going to bed, and in this way he obtained very good rest at night; and with the help of a simple acid and bitter tonic recovered his health completely. The urine of this patient was examined twelve, and also twenty hours after the dose of æther was taken, and found to contain no sugar.

Experiment XXVIII.—A gentleman who came under my care for several months was a sufferer from spasmodic asthma, uncomplicated by any recognizable organic disease of heart or lungs. He was 37 years old, and was occupied chiefly in literary work, at which he invariably labored very hard : the malady was of twelve months' standing when I first saw him, and the attacks had become very frequent, they would return night after night for a week or ten days, and then leave him comparatively free during several days. This gentleman had contracted the habit of easing his sufferings by the inhalation of æther from a handkerchief; he never carried the process so far as to produce any true symptom of anæsthesia, but contented himself with frequent whiffs, which just sufficed to give considerable relief for a time, to the severe dyspnea. The manner of life of this patent, during the periods when he was suffering regular exacerbations every night, was very singular ; he may be said to have lived in an atmosphere of æther vapor, for not only did he actually absorb considerable

quantities, in repeated small doses, by the lungs, but, owing
to the wasteful manner in which the inhalations were conducted, the air of his bedchamber was always heavily charged
with it, more especially during the night, when his sufferings
were most distressing.   Yet there was never an approach to narcosis in any true sense of the word ; and it is an interesting fact,
that in spite of the repeated action of æther upon the organism the urine never contained more than a very small trace of
sugar, which was equally present at times when no æther had
been used for many days, and which I believe was due to
the malady and not in any sense to the treatment.   Yet the
same dose of æther inhaled by a healthy person, and so frequently repeated, would almost certainly have caused a very
considerable diabetes.   Another remarkable feature in the
case was the extraordinary independence of common food displayed ; notwithstanding the severe fatigue inflicted on him
during the times when the exacerbations occurred nightly,
the patient, who was rather a spare man, ate an extremely
small quantity of food, and of this almost none in the solid
form.   Yet at the end of one of the severest bouts which I witnessed, on his somewhat suddenly losing the attacks, he really
appeared scarcely the worse for what had passed ; and even
allowing for his enforced confinement strictly to one room, it
was surprising to observe to what a trifling extent emaciation had occurred.   The result, if due to the æther at all,
was probably due in great part also to a daily ration of
something like a half bottle of port wine.

B.—Effects on Animals.

The observations which I have made in this direction are
confined to one point, with regard to which, however, some
important facts have been elicited ; namely, the influence
which æther exerts on the affections of the nervous system
caused by certain poisons.   It occurred to me that it would
be especially interesting to investigate the action of æther

in strychnia poisoning. Preliminary trials established the fact that $\frac{1}{1000}$ grain of strychnia injected into the cavity of a frog's body induced tetanic spasms, on the average, within two minutes. Other preliminary experiments satisfied me that frogs, placed in an atmosphere equally charged with the vapor of one grain of æther to each 100 cubic inches, became affected with the first decided symptoms of narcosis in from five to six minutes.

Experiment XXIX.—A large glass jar being charged with the above-mentioned proportion of æther vapor, a healthy frog was placed in it, into the cavity of whose body $\frac{1}{1000}$ grain of strychnia, dissolved in five minims of distilled water, had that moment been injected. No symptoms indicative of strychnia poisoning manifested themselves until the animal had come decidedly under the narcotic influence of æther (viz. in about five minutes), so far under it at least as was indicated by complete palsy of sensation in the hind limbs. Soon after this a tremulous motion of the head was noticed, and the hind limbs, which had evidently by this time become palsied, were gradually stretched out to their full length, and became agitated with successive shocks, as it were, of tonic spasms. No very high degree of tetanus was induced; and the animal survived for twenty-five minutes, dying, to all appearance, of æther-narcosis, but suffering from frequently recurring though slight tetanic spasms throughout. Rigor mortis immediately set in.

There could be no doubt that the narcotic influence of æther delayed the fatal termination of strychnia poisoning here—but how did it do this? Simply, as I believe, by moderating the violence of the spasms which would have exhausted the animal; this it probably effected by destroying more completely the vital communications between the muscles and the nervous centers, and thus inducing a variety of narcosis in which paralysis naturally predominated, rather than convulsions. The delay in the *first occurrence* of the tetanic symptoms was, I believe, owing to an entirely different cause : to the stimulant action, namely, of the small quantity of æther which alone had been absorbed at that early stage, and which had not sufficed to produce any symptoms of narcosis; for in the following experiment it was shown that a non-narcotic dose of æther *may* produce such an effect.

Experiment XXX.—$\frac{1}{1000}$ of a grain of strychnia, dissolved in half a minim of æther and five minims of distilled water, was injected into the cavity of a frog's belly. No unconsciousness, nor paralysis, nor any tetanic symptoms were developed for at least ten minutes; at the end of this period the ordinary symptoms of strychnia poisoning began to appear, but they were not fully developed at first, and the animal lived eight minutes longer, preserving voluntary power and consciousness (in the interval of the spasms) to the last. Rigor mortis immediately followed death. Here it was plain that a dose of æther insufficient to narcotize suspended the action of strychnia; and it is equally obvious, in comparing this with the preceding experiment,* that it was not in virtue of any *direct antidotal action* that æther counteracted for a time the fatal influence of the other poison. In both cases there was the stimulant action of the small dose at first, which so acted beneficially as long as it lasted. And in the first experiment there was a further narcotic effect (when a large dose of æther had been absorbed), which, by destroying more rapidly the efficiency of the spinal motor nerves as communicators of nervous influence, kept the muscles in comparative quiescence, and thus delayed fatal exhaustion. Meantime the strychnia did not at all interfere with the action of the æther.

These experiments were afterward repeated, with suitable variations of dose, on rabbits and on rats, and the results arrived at were similar; but there is not space here for details. Like results were also obtained, when the aqueous solution of opium (in commensurate dose) was substituted for strychnia, in frogs.

---

* Experiments XXIX. and XXX. were each repeated three times in frogs, with substantially the same result.

Enough evidence has now, I think, been adduced to establish a very complete distinction between the stimulant and the narcotic effects of æther, and to show that they have only that accidental relation to each other which consists in the fact that the one kind of influence is produced by small and the other by large doses, that they are never truly intermixed, but that the latter may follow the former in a merely accidental sequence, during the process of absorption of a large dose.

And it has also been shown that a number of phenomena which have been improperly referred to the stimulant action of æther (phenomena of "excitement") are really a part of its narcotic action.

CHLOROFORM.

RESEARCHES ON THE PHYSIOLOGICAL ACTION OF CHLOROFORM.

THIS inquiry was directed to three objects :—I. The complete investigation of the phenomena of non-fatal chloroform-narcosis ;—II. Examination of the phenomena of fatal chloroform-narcosis, and ;—III. Observations on the stimulant action of chloroform.

I shall make no apology for treating this subject *de novo*. My justification for this must be that peculiar circumstances allowed me, during a period of six years (from 1855 to 1861) to administer chloroform in one way or another 3,058 times: the large majority of the patients being hospital inmates, in whose cases there were facilities for preserving an accurate record of all the important features of an inhalation. Having made this general statement, I must be asked to be excused from making distinct reference to many observations, by other writers, which I should otherwise feel bound to discuss; since my space would otherwise be insufficient.

20

PART I.

The Phenomena of non-fatal Chloroform-Narcosis.

THE following are the subdivisions of this branch of the investigation:

1. The phenomena of the successful induction of chloroform-narcosis in man.

2. The parallel phenomena in animals.

1. Before commencing particular observations on the symptoms of chloroform-narcosis in human beings, it was necessary, by preliminary trials, to satisfy myself as to the conditions which ought to be observed, in order that the experiments might be decisive. In the first place, it was determined that if uniform results were to be obtained, it must be by constantly employing an instrument which would at least approximately regulate the course of the administration. Numerous trials convinced me that of the then existing apparatus for the purpose, the inhaler of Snow yielded the most regular results. On the average being taken of the periods of time consumed in inducing full anæsthesia in above 300 cases, it was found that *four minutes* was the time occupied in those cases which were distinguished by the greatest regularity in the progress of the symptoms. Observation was then recommenced; the phenomena of as large a number of cases as possible were noted; but, from the total number, only the reports of those in which the induction of anæsthesia occupied four minutes, or within a few seconds more or less of this period, were preserved. When the num-

ber of such cases included fifty children and fifty adults, this
part of the observations was closed, and this select group of
cases **was** analyzed. It must be understood that by "children," are signified patients under **fourteen** years of age ; by
"adults," patients more than fourteen **years old.** It should
be further observed that of these hundred patients, none
were the subjects, so far as could be discovered, of disease of
the heart, lungs, or brain.

The points which were specially investigated were—A—
the effects on the pulse; B—the effects on respiration; C—
the effects on the general nervous system.

A. The pulse. First, with regard to the fifty adult patients (of whom twenty-four were males and twenty-six
females). The pulse was quickened during the first two or
three minutes in forty-eight ; in forty of these the *force* of
the heart's action was also manifestly increased at first,
though this soon fell off : this was tested not merely by the
pulse, but also by stethoscopic examination of the heart. Of
the eight patients in whom the heart's action was quickened
without increase of its force, **six** were women and two men :
in three of the women and in one of the men the pulse was
irregular, and occasionally intermitting. In the two patients
in whom no quickening of the pulse occurred, there was a
remarkable increase in the force of the heart's action ; the
pulse was strong and full : the cardiac impulse strong and
heaving ; the systolic sound long and loud ; the diastolic not
particularly altered ; this increase of power was maintained
throughout the whole period of anæsthesia. Both these
patients were males, their ages 25 and 28 respectively ;
neither of them were at all convulsed.

In the fifty children, it was not attempted to make systematic observations of the pulse-rate during the *early* stages of
chloroformization, since the influence of *terror* would, in most
instances, have falsified the results.

In both adults and children, it was noted that with the
advent of the "third stage" of anæsthesia (contracted pupils,

loss of consciousness, but not of muscular irritability), the pulse sank to nearly its normal frequency, and remained at or below that level during the whole continuance of anæsthesia, provided the case progressed favorably. Sustained increase of the pulse-rate after the induction of the third stage was noted in four patients, three of whom were adult females, and one an adult male. It was noted that the change from the horizontal to the sitting posture had far less influence on the pulse-rate than in ordinary circumstances. Besides the cases in which the increased pulse-rate was maintained during the third stages of anæsthesia, momentary quickening of the pulse occurred in the period of deep unconsciousness in fifteen patients, eleven of whom were adult females and four were adult males: in all these cases it was accompanied by marked failure in the power of the heart. In twelve out of the fifteen cases, the momentary quickening took the form of a quick, unrhythmical running; and in one patient, a man undergoing a severe and bloody operation for the removal of the upper jaw, extreme pallor of the face and gasping respiration occurred.

The subjoined table represents the effects produced upon the pulse of adults during the whole period of the induction of anæsthesia, and the state of the pulse in children during the later stages only. The words "average excess" are intended to imply the excess in frequency over the pulse-rate noted *at the commencement of inhalation*, and where, instead of an excess, a diminution was observed, this is signified by prefixing the *minus* sign.

The table of observations on adults must be taken subject to the following allowances: 1. for the influence of *terror* in quickening the pulse; 2. for the fact that in twelve cases anæsthesia was induced in from twenty to thirty seconds earlier than the fourth minute, and in three cases its advent was delayed a few seconds beyond that period.

The table of observations on children does not pretend to record the state of the pulse earlier than the end of the second minute from the commencement of inhalation, by which time the influence of alarm is obliterated by the failure of consciousness.

Observations were made at the end of the first, second, third, and fourth minutes respectively.

	Subjects.	Average excess 1st minute.	Average excess 2d minute.	Average excess 3d minute.	Average excess 4th minute.
50	Adults.				
24	Males	20·0	18·5	8·5	0·4
26	Females . . .	26·4	15·5	9·2	4·38
50	Children.				
32	Males		10·45	7·14	—0·4
18	Females . . .		14·2	7·9	—0·5

A review of the above table appears to warrant the following conclusions: 1. In the successful and regular induction of surgical anæsthesia by chloroform, the pulse is very considerably increased in force and frequency during the first minute of inhalation, under the effects of chloroform, or of this, combined with the patient's alarm. 2. At the end of the second minute (the effect of alarm being done away with), we still **find** a considerable increase of the frequency of the pulse. 3. **At** the end of the third minute, we find that the pulse is rapidly dropping to its previous level, under the influence of chloroform. 4. At the end of the fourth minute, complete anæsthesia being induced, the pulse is commonly reduced nearly to, or even below, its normal level.

B. Respiration was carefully observed in the fifty adult patients, and the subjoined table was drawn up. It is obvious that the quick breathing rate at the commencement arose from alarm; the average rate was much higher in women than in men; occasionally, in a very nervous female, it rose **as high as** one might expect to find in pleurisy. It is to be **noted that in** fourteen cases some temporary hurry or irreg-

ularity of respiration was noted during the period of complete anæsthesia (fifth to tenth minute inclusive). All these
cases were attended with disturbance of the pulse, which was
either rapid, or feeble, or intermittent, or all these at once.
In four of these cases, respiration assumed a *gasping* character for the time, and in one of them (the case of removal of
the jaw already referred to) this feature was developed to an
alarming extent.

	Subjects.	Average frequency at commencement	Average frequency end of 2d minute.	Average frequency end of 4th minute.	Average frequency end of 6th minute.	Average frequency end of 10th minute
	Adults.					
24	Men . . .	20·8	18	16	14·8	14·2
26	Women. .	26	19	16	15·5	15·5

The following conclusions seem warranted. 1. Chloroform,
by the operations of its smaller doses, calms the excited respiration of alarm, or physical feebleness.* 2. Beyond this,
it has no further effect on respiration, unless anæsthesia is
carried to a dangerous degree, or, at most, only a slightly
depressing effect in the later stages. 3. Irregularity, marked
slowness, or a gasping character of breathing, are to be considered as signs of an irregular or excessive administration of
the anæsthetic.

C. The rest of the phenomena noted are those plainly
traceable to alterations in various portions of the nervous
system, and include all full report as to the adult patient,
and an equally complete account of the symptoms observed
in the children, except as to the phenomena of consciousness,
which it would not have been possible to report with accuracy.

With regard to the fifty adults, the subjoined table records
observations made at the end of the first, second, third, and
fourth minutes respectively. At each of these periods a
record is preserved of the number of cases in which phenomena, under the several headings, were observed.

* This is no part of the really narcotic effects.

This table exhibits at a glance the order in which the various nervous functions are affected by its arrangement from left to right. It remains to make some observations on the phenomena included in each column:

	Effects on Intellect.	Effects on Voluntary Co-ordination.	Effects on Sensation and Voluntary Motion.	Effects on Sympathetic System.	Effects on Medulla Oblongata.	"Reflex" Phenomena.
End of						
1st minute	42	17	4	10	...	10
2d minute	50	46	47	21	...	16
3d minute	50	50	50	42	3	28
4th minute	50	50	50	48	12	9

1. The effect upon the intellectual operations during the few first seconds of inhalation appears to be an invigorating one. Needless timidity and distress vanish, while consciousness is as yet unaffected, and the expression of the face is bright and intelligent. Before the conclusion of the first minute, however, the approach of intellectual paralysis is perceived, in the great majority of cases; the patient, if narrowly observed, will be found to be losing the consciousness of actual events; and in many instances this state is accompanied by a disposition to voluble talkings of a more or less unreasonable character. As the affection increases, the very curious phenomena of narcotic reminiscence, already alluded to in the section on Narcosis, make their appearance (if at all): this never happens till inhalation has proceeded for at least a minute, and usually longer. Paralysis of the tongue usually prevents the utterance of connected language in a few seconds more, but the patient will sometimes continue to babble in a way indicating the existence of consciousness till a very advanced stage of anæsthesia; and in cases where it is not necessary to carry narcotism to its furthest stage, this may go on throughout the maintenance of anæsthesia. Signs of emotional disturbance sometimes accompany the first

diminution of consciousness; the patient is angry, or pathetic, or sentimental, and gives expressions to these feelings in words, with no reference, or only a confused reference, to persons and things actually present.

2. The power to co-ordinate voluntary movement is usually affected very soon after the first impairment of consciousness; the patient makes vain efforts to grasp objects, and soon afterward his articulation becomes slightly impaired.

3. Sensation and voluntary motion. These functions, although their alteration may be said on the whole to proceed *pari passu*, both of them being earliest disturbed in the *lower limbs* (as was observed of alcohol), must be separately reported on.

a. The first effect on sensation which is noticeable is the removal or mitigation of any pain from which the patient suffers; this is often effected by one or two inspirations only. Evidence of *paralysis of skin sensibility*, on the other hand, could only be obtained in four patients during the first minute, and only in the lower limbs. Before the end of the second minute, however, there was very considerable paralysis of the whole skin surface in forty-seven out of the fifty patients. The conjunctiva always retained sensibility later than the skin, with certain exceptions, presently to be noted; in the great majority of instances, however, it was rendered insensitive by the time that contraction of the pupil was well marked (third stage). Certain portions of the skin and subcutaneous tissue, however, retain their sensibility with extraordinary tenacity: these are the matrix of the great toe-nail, the margin of the anus, and the whole of the skin of the organs of generation. It is impossible to obliterate their sensibility without pushing chloroformization to a degree which greatly surpasses that required for ordinary purposes. This observation is confirmed by my experience with animals, and its importance cannot be too highly estimated; for it explains the frequency with which death has happened in the course of anæsthesia induced for the performance of operations

for phimosis, evulsion of the toe-nail, hæmorrhoids, &c. All
kinds of fanciful reasons have been given for this fatality of
chloroform in such trifling operations; but there is no doubt
in my mind that this is the true one.

β. **Voluntary motor power is usually** not much diminished
during the first minute of inhalation, and the patient occasionally struggles **from** alarm; but in the case of *regularly* induced anæsthesia, this voluntary struggling is limited, in the
great majority of instances, to this early stage, and simple
paralysis, gradually increasing, is the only further phenomenon connected with the voluntary muscles. This paralysis
had reached a high degree by the end of the second minute,
in thirty-six out of the fifty patients.

4. Under the head of "effects on the sympathetic system,"
are classed—(a) Flushing of the face, accompanied with heat
(and sometimes with sweating), and not fairly attributable to
the heat of **the room** or to the effect of struggling, &c.   This
was observed in nine cases, and *was always a late phenomenon,* unlike the case of alcohol and of **æther, in** which sympathetic paralysis is one of the earliest symptoms. (β) Contraction and dilatation of the pupil. The changes in the pupil are
recorded in the subjoined table :—

	Pupil contracted.	Pupil dilated.
End of 1st minute . .	4	
End of 2d minute . .	42	2
End of 3d minute . .	43	3
End of 4th minute . .	34	9

In **seven patients** it was impossible to form a fair judgment
of the state of the pupil, but in forty-three the changes were
evident; and while contraction occurred in the majority of
cases by the end of the second minute, and was fully developed
during the third minute, dilatation rarely occurred till the
latest stages, and may fairly be considered as a sign that a
profound degree of anæsthesia has probably been reached.

This observation has been verified by many cases besides those here tabulated. Wide dilatation of the pupil is very rarely witnessed, except in conjunction with stertorous breathing and considerable depression of the heart's action, both as to force and frequency.

5. By effects on the medulla oblongata are meant the production of stertorous breathing, laborious breathing, or both. The former was noted in three cases as early as the end of the third minute, in nine others, not occurring till near the end of the fourth minute; the phrase is by no means intended to signify mere *snoring*, but paralysis of the velum palati and buccinators. Laborious breathing (*i.e.* slow and somewhat irregular) was only noted in three cases; it commenced just before the induction of full anæsthesia. And in one of these (already referred to), during the performance of the operation and immediately after a great loss of blood, the breathing suddenly became hurried and shallow, and then changed to a series of gasps. Profound modification of the breathing was always accompanied by wide dilatation of the pupil.

6. "Reflex" phenomena. *a.* .Coughing was produced in ten cases during the first minute, obviously from irritation of the air-passages by the vapor; in six others not till during the second minute. It did not occur in any patient after the conclusion of the second minute, except as a sequel of vomiting. *β.* Vomiting occurred four times during the second minute; never earlier than this. (It must be noted that all the patients had abstained from food for a considerable time previous to inhalation.) It occurred in eight patients during the third minute, and in three patients either occurred or was first produced during the fourth minute. Vomiting was only noticed in two cases after the full formation of the anæsthetic state; in one of these it happened during the reaction from a state of partial syncope. *γ.* In two cases involuntary expulsion of fæces took place; this happened during a state of complete anæsthesia, and in both instances was associated with marked flagging of the pulse and considerable pallor of countenance.

The impression left on my mind by observation of the reflex phenomena above mentioned is that, with the exception
of coughing produced by the irritating effects of the vapor on
the air-passages, they are symptoms demanding careful attention.   I believe that vomiting, and more especially involuntary expulsion of fæces, occurring during the advanced stages
of anæsthesia, unless they are caused by considerable loading
of stomach or rectum, represent a very considerable depression of the nervous system, in which the occurrence of syncope
is to be especially guarded against.   In one fatal accident
from chloroform which I witnessed, though it fortunately did
not happen under my hands, the urine was forcibly ejected
almost immediately before death ; **and I have often** observed
the same occurrence in animals.   When, however, vomiting
occurs *after* a state of syncope it is to be looked on as a decidedly favorable symptom, indicating a certain amount of
reaction.

   ð. Spasmodic rigidity of the principal flexor muscles was
noted in five cases during the second minute, and in twenty-
six cases during the third or **fourth minute.**   It was in all
instances *symmetrical*, affecting both sides of the body alike ;
in ten cases it was accompanied by barely perceptible clonic
convulsions ; and when this was the case the rigidity of the
neck muscles was very great, and there was momentary sus-
pension of the respiration, and consequent congestion of the
face.   Of the five patients in whom rigidity of the muscles
occurred *early*, four were females and one a weakly young
man.   Of the twenty-six patients in whom it occurred later,
eight were females and eighteen were males.   Of the ten
cases in which slight clonic convulsions were noted, seven were
females and three males.

All these convulsive phenomena were very transitory, and
in no case was the inhalation entirely suspended.

In the fifty children, observations were made on the modifications of sensation and voluntary motion, of the functions
of the sympathetic system **and** the medulla oblongata, and on

the "reflex" phenomena. The only matter requiring observation is, that convulsive phenomena were noted in only two cases, and symptoms indicating approaching syncope *only in one*. The latter fact faithfully reflects the result of my whole chloroform experience, in demonstrating the comparatively small amount of danger to *children* from weakening of the heart's action under chloroform.

The final observation which has to be recorded as to these 100 cases is, that the smallest quantity of chloroform employed in any case was two, and the largest three, drachms.

On a review of the observations detailed above, I conclude that—

I. The phenomena of anæsthetization of man by chloroform vapor, when this is conducted with method, and is accomplished in a period of about four minutes, are, on the whole, extremely regular.

II. Under such circumstances the occurrence of any dangerous degree of depression is rare, and may always be remedied by attention.

III. During a very short period at the commencement of inhalation the vital functions are stimulated, an effect which ceases on the very first occurrence of narcotic symptoms.

IV. The whole subsequent train of symptoms is that of depression of the various portions of the nervous system in consecutive order of occurrence; the brain being the first to lose, and the medulla oblongata and sympathetic system the most tenacious in retaining, their proper functions.

V. The convulsive phenomena occasionally witnessed belong to the category of the paralytic, and not of the stimulant effects of chloroform, unless when they are brought about by an accidental local cause.

VI. Every considerable increase or decrease in the frequency, and every decrease in the force, of the respiratory and circulatory movements is a part of the depressing influence of chloroform.

Narcosis from Chloroform ingested by the Stomach.

As an appendage to the above observations, I shall add the report of some researches made by me on narcosis in the human being, produced by chloroform taken into the stomach. The very possibility of such an occurrence has been denied by some observers, and I have therefore thought that these facts might be interesting.

Passing over some preliminary trials with increasing doses, which were not sufficiently large to produce the desired effect, I shall detail the result of an experiment which sufficiently demonstrated the possibility of anæsthesia being produced by the ingestion of chloroform by the stomach.

Forty-five minims of chloroform, suspended in an ounce and a half of thin. mucilage, was swallowed, the stomach being quite empty at the time. Great warmth at the epigastrium, and a feeling of flushing all over the body succeeded, almost at once; five minutes after taking the dose, the pulse was throbbing 100 per minute, and the heart beating with uncomfortable violence; a sense of decided confusion of mind also annoyed me. Five minutes later I experienced a considerable degree of nausea, and the pulse had fallen much lower, but it is impossible for me to speak to its positive frequency, as I must have fallen very soon after this into a state of unconsciousness. I recovered my senses at length, and on looking at my watch found that it was forty-six minutes from the time of commencing the experiment. That it was not common sleep into which I had fallen was obvious from the fact that my lower limbs still felt heavy and numb, and on attempting to stand I tumbled down. For almost two hours after this I remained in a state of great discomfort, shivering, nauseated, and with aching pains in the head and in all my limbs, which sometimes assumed the sharpness of a twinge of neuralgia. It was some time, also, before I recovered my muscular sense and an accurate co-ordinating power over the movements of the limbs.

The result of this experiment surprised me, as at that time I supposed that anæsthetic effects were never produced by this mode of administration of chloroform. This idea, however, was incorrect, for in a case recorded by Taylor* one drachm swallowed produced fatal coma in a child; and (oddly enough) Messrs. Lallemand, Duroy, and Perrin, in the very work in which they deny the anæsthetic action of liquid chloroform, relate a case in which the same dose very nearly proved fatal to an adult man. The recovery was excessively slow in this case, which occurred in England, and was originally reported in the *Medical Gazette.*

2. *Complete Anæsthesia in Animals.*

A. Experiments on the Effects of Inhalation.

In the following experiments great care was taken to regulate accurately the proportion of chloroform vapor in the inspired air. The jars, also, in which the animals were placed were always of so large a size as to render it impossible that deprivation of oyxgen should make any considerable addition to the injuries inflicted by the anæsthetic.

Experiment I.—A large glass jar was charged with two per cent. of chloroform vapor, and a full-grown sparrow was introduced into it.

In one minute the bird began to reel as if tipsy, and it soon fell on its side. At the end of three minutes it was removed to the outer air and found to be wholly insensible: the breathing was weak and flurried; heart acting strongly and regularly. After two or three minutes' exposure to the air it was found, first, that the conjunctiva, and then that the wings were sensitive, and in a little time consciousness was fully restored. In about six minutes from its removal to the air the bird was able to fly pretty steadily.

Experiment II.—The same jar was charged with two per

* On Poisons, p. 740.

cent. of chloroform vapor, and a young rabbit was introduced. In one minute and a half the eyes closed, and the breathing, which had been hurried from alarm, became tranquil; at the end of four minutes it was removed from the jar: it was then observed that there was considerable rigidity of the posterior limbs, and occasional convulsive shuddering of the whole body. The heart's action was steady and quiet. After five minutes' exposure to the fresh air the animal began to recover itself, and moved feebly, dragging its posterior limbs. It soon recovered completely.

Experiment III.—A full-grown sparrow was introduced into the jar, which was charged with two per cent. of chloroform vapor, and allowed to remain there five minutes. The same phenomena occurred as in Experiment I.; but on removing the bird from the jar, respiration was found to be slow and gasping, and the heart's action very hurried and weak. Recovery took place in twelve minutes.

Experiment IV.—The jar was charged with two per cent. of chloroform vapor, and a young rabbit introduced. In two minutes, he was lying on his side, completely paralyzed as to his limbs, but not quite unconscious, for on shaking the jar he lifted his head. At the end of three minutes he was completely comatose. At the end of four minutes he was removed to the outer air; respiration was regular, and not particularly rapid; heart's action rather slow: there were slight twitching movements of the hind limbs. Recovery took place in ten minutes, sensation in motor power returning earliest in the fore limbs.

Experiment V.—A small terrier was made to breathe an atmosphere of two per cent. chloroform vapor, stored up in a very large bladder, and which he was made to respire over and over again, the bladder being connected by a caoutchouc tube with a smaller one, which inclosed his head, and was drawn close by a running string round his neck. After a few moment's struggling, the animal submitted quietly enough, and the inhalation proceeded regularly. At the end of a minute and a half, the hind quarters were found to be paralyzed as to voluntary motion and sensibility; but the skin of the neck, the fore paws, and the tips of the ears, which were left uncovered, were still sensitive. At the end of three minutes' inhalation, the animal rolled on its side perfectly paralyzed and quite unconscious. Respiration 28, circulation 72. The bladder was removed, and in five minutes the animal was quite conscious, but still completely paralyzed as to its hind limbs, and partially so as to the fore limbs. The conjunctiva recovered its sensibility earlier than any part of the skin, except the ano-genital region, which had never quite lost it. The power of co-ordinating the movements was not completely restored for an hour.

Experiment VI.—A healthy terrier, weighing 8 lbs. 2 ozs., was placed in a jar which was charged with 4·5 per cent. of chloroform vapor. It was withdrawn at the end of four minutes, completely unconscious, the limbs relaxed, respiration 12 per minute and gasping; heart's action weak, 120. Removed to the fresh air, the dog immediately drew several forcible breaths, and the strength of the pulse increased. Five minutes later the animal was conscious, though still drowsy, and almost completely paralyzed in its hind quarters, both as to sensation and voluntary motion. The ano-genital region recovered sensibility *first*.

Experiment VII.—A full-grown rat was placed in a jar which was charged with an atmosphere of two per cent. chloroform. It was removed after three minutes, and was then profoundly unconscious, skin everywhere insensitive; conjunctiva insensitive, eyeballs moving outward and inward alternately, in a convulsive manner; posterior limbs slightly convulsed; breathing and circulation calm. Recovery took place in ten minutes.

Experiment VIII.—A large and active frog was placed in a jar which was charged with 0·5 per cent. of chloroform vapor. Full anæsthesia and unconsciousness was induced in

21

twelve minutes, but the conjunctiva retained its sensibility. Recovery took place in twenty minutes.

Experiment IX.—A large and active frog was introduced into a jar which was charged with one per cent. of chloroform vapor. In seven minutes the animal was completely anæsthetized, and the respiration was interrupted, though the heart continued to act. Recovery occupied about fifty minutes, the posterior extremities recovering motion and sensibility last. During the first few minutes after removal from the jar, the anterior limbs assumed a *rigid* state, which did not relax for some little time.

B.—Injections into the peritoneal cavity.

Experiment X.—A very large and strong rat, which had been fed plentifully six hours previously, had ten minims of chloroform injected into its peritoneal cavity at 12.15 P.M.

The sole effects of this injection were the production of drowsiness and partial paralysis of sensation and motion in the hind limbs. One hour from the time of injection the animal had apparently quite recovered.

At 1.18 ten minims more were injected; this time a greater effect was produced, for in fifteen minutes there was complete paralysis of both hind legs and of the right fore leg. The animal, however, still retained a good deal of consciousness, and struggled with its head and its left fore leg when touched. Complete paralysis, however, did not occur at all, nor complete unconsciousness. In the course of an hour and forty minutes it had entirely recovered.

The chloroform used in this experiment was examined and tested, and was ascertained to be a very poor specimen. It was therefore laid aside.

Experiment XI.—Ten minims of chloroform (Jacob Bell's) were injected into the peritoneal cavity of a full-grown rat, of remarkable size, at 11.48 A.M. The animal had been fed at nine o'clock.

Complete anæsthesia was induced in five minutes, and death (by cessation of respiration) in thirty-nine minutes, from the time of injection. It was evidently necessary to use smaller doses to avoid producing fatal anæsthesia.

Experiment XII.—Five minims of chloroform (Bell) were injected into the peritoneal cavity of a large full-grown rat at 2 P.M. (The rat had been fed at nine o'clock.)

2.5. Paralysis of sensation and voluntary motion in hind quarters; breathing regular and calm.

2.10. Complete unconsciousness, with relaxation of all the limbs; insensitiveness of all the skin and of the conjunctiva of the eye. Breathing regular and quiet.

2.23. Retrocession of the symptoms is commencing; consciousness has partly returned.

2.42. The animal got on its feet and ran nimbly; recovery was perfect. (No evil consequences appeared to result from the damage to the peritoneum either in this case or in Experiment X.)

Experiment XIII.—A full-grown guinea pig had twenty minims of chloroform (Bell) injected into the peritoneal cavity at 1.30 P.M. with a Wood's syringe. Ten seconds after the withdrawal of the syringe the animal gave a cry, and fell on its side, the limbs extended and rigid. Fifty seconds later clonic convulsions of the hind limbs occurred, which lasted a few seconds, and were succeeded by paralysis of the same.

At 1.32 the animal was lying deeply comatose, limbs relaxed, general anæsthesia, except in the ears, the ano-genital region, and the conjunctiva.

1.35. The animal exhibited more sensibility in the anterior limbs, and when touched moved them feebly.

Recovery from anæsthesia was complete in an hour and a half, but symptoms of peritonitis set in, and the animal died in the night. Post-mortem examination showed deep congestion of the peritoneal covering of the bowels in many places; some liquid chloroform (detected by its smell) mixed with a little bloody serum was found in the belly.

Experiment XIV.—A full-grown cat had thirty minims of chloroform (Bell) injected into the peritoneal cavity at 1 P.M. Complete anæsthesia was induced in about ten minutes, with the regular train of symptoms. Recovery commenced at 1.48. At 2.30 consciousness was complete, but there was still palsy of the hind legs. On the following day the animal was in full health. No bad symptoms supervened.

The special conclusion drawn from these experiments upon peritoneal injection was, that this method of introducing chloroform to the system is available as a standard of comparison with the effects of the anæsthetic as inhaled by the lungs. Various conditions with which I was not then, nor am now, sufficiently acquainted, affect the rapidity with which chloroform is absorbed through the peritoneal membrane, and these deserve a special investigation. It is enough for our present purpose that the anæsthetic, either in liquid or in the form of vapor, is rapidly absorbed into the blood, as indicated by its physiological effects, and that these effects correspond in degree to the amount of the dose injected. The latter point will be rendered more evident in our description of the phenomena of fatal chloroform-narcosis.

From the whole series of experiments on animals above detailed, we are warranted in the general conclusion, that the action of chloroform in narcotic doses insufficient to destroy life is essentially the same, as regards the effects upon the nervous system, in the case of mammalia generally as in the case of man ; and that with regard to birds on the one hand, and frogs on the other, the differences in the phenomena observed may be readily accounted for by the extreme activity of respiration in the first, and the absence of true lung-respiration and the low temperature in the second.

There is one result of the narcotic action of chloroform which must not be passed over in silence, though I am not able to supply such accurate information with regard to it as has been given on other points—viz. the production of an artificial diabetes analogous to that which has been described as caused by æther.

Dr. Pavy* states that, of twenty patients whose urine he examined shortly after their recovery from chloroform-narcosis, only one failed to present evidence of a decided diabetic condition, the result of the action of the chloroform. For my own part, as regards anæsthesia by inhalation, I can only speak generally to the fact that the production of diabetes is a very frequent, if not constant, result of full anæsthesia by inhalation. I am not prepared to say how much of this depends on mere diminution of the respiratory movements, as suggested by Reynoso,† and how much upon a paralytic affection of the sympathetic nerves distributed to the liver, such as I believe to be the probable cause of diabetes following the ingestion of æther by the alimentary canal or portal vein. But this is certain, that diabetes is easily produced by the injection of chloroform, not only into the portal vein, but also into the peritoneal cavity—that is, if so large a dose be not given as to prove fatal before the sugar formation can attain the degree which is necessary in order that it may overflow by the kidneys. In this case I feel convinced that the effect produced is one simply of palsy of the sympathetic branches in the liver, and I suspect that even in anæsthesia by inhalation this is at least the principal source of the affection, inasmuch as paralysis of the other parts of the sympathetic system occurs simultaneously.

* Op. cit.
† Bulletin de l'Académie de Médecine, tom. xxxiii.

PART II.

Phenomena of Fatal Chloroform-Narcosis.

1. *Experiments on Animals.*

A.—Death caused by comparatively moderate doses.

EXPERIMENT XV.—A rather small, but apparently full-grown rat had ten minims of chloroform thrown into its peritoneal cavity at 12 noon.

Fifty seconds later the animal was very restless; the heart's action and breathing extremely rapid.

12.1.20. The rat was seized with slight clonic convulsions of the posterior limbs, which lasted a few seconds; when these had ceased it was found that the animal was quite conscious, but dragged the hind limbs, in which sensation was obviously blunted.

The deepest degree of anæsthesia was reached at 12.3, the animal being at that time unconscious, heart's action and breathing regular; sensibility, however, still existed in the face and the conjunctiva, and, to a slight degree, in the anterior extremities.

12.4.30. As anæsthesia was rapidly diminishing, ten minims additional were injected. A minute later, the heart's action was found to be excessively rapid and weak, respiration flurried and almost entirely abdominal.

At 12.6 respiration ceased. The thorax was quickly opened, and the heart was seen feebly pulsating; this continued for seven minutes; and for several minutes after its cessation, contractions could be re-excited by irritants.

Experiment XVI.—A full-grown rat was introduced into a jar which was charged with two per cent. of chloroform vapor. Symptoms of anæsthesia were produced in less than a minute. Respiration ceased at the end of ten minutes. The thorax was now immediately opened; the heart was seen to beat for eight minutes longer, and even then continued irritable for some time.

Experiment XVII.—A full-grown sparrow was introduced into a large jar charged with two per cent. of chloroform vapor. At the end of eight minutes, respiration being very slow and gasping, the bird was removed to the outer air, but breathing ceased in less than two minutes. The heart continued to beat for nearly twenty minutes.

Experiment XVIII.—A dog, full-grown, weighing 14 lbs. 8 oz., had ꝫii of chloroform (Bell's) injected into his peritoneal cavity, at 1.30.

Paralysis of the posterior limbs was induced in about eighty seconds.

1.33. The animal lay in a state of profound anæsthesia, every portion of the skin insensitive, except the ano-gential region. Conjunctiva insensitive; pupils dilated and fixed; globes of the eyes convulsively rotated by alternate contractions of the internal and external recti.

1.42. Respiration had become very much hurried and agitated, the animal, though deeply unconscious, utters cries, and there are slight convulsive movements of the hind limbs. Pulse 100, nearly regular.

1.52. Profound and tranquil anæsthesia. Respiration 45, regular. Pulse 84, steady and tolerably strong.

About 2.30, the respiration became first very rapid, and then slow and gasping; and finally it ceased, exactly at three o'clock. The chest was immediately opened, and the heart was found pulsating strongly. In less than ten minutes the pulsations were confined to the auricles, but a prick with the knife-point reinduced action of the ventricle for another minute or so; this was repeated several times; and when (at twenty minutes from death) the ventricle was no longer irritable, rough friction of the thoracic ganglia of the sympathetic repeatedly reinduced strong contractions.

Experiment XIX.—A full-grown cat had ʒi of Bell's chloroform injected into the peritoneal cavity at 1.30 P.M.

Three minutes later the hind legs were found perfectly paralyzed as to sensation and voluntary motion; the pupils widely dilated, but still slightly sensitive. At this moment the urine was convulsively discharged, and at the same time the animal passed, rather suddenly, into a state of coma, from which nothing could rouse it.

1.45. There is most profound insensitiveness everywhere of the skin; wide and fixed dilatation of the pupils. The eyeballs have the same convulsive motion noticed in the dog in Experiment XVIII.; the tongue is convulsively protruded and withdrawn many times in a minute. Respiration 60; pulse 100.

2.2. Respiration 60, regular. Heart's action a mere uncountable run; very weak.

2.5. Clonic convulsions of the right fore and hind leg; foam at the mouth.

2.10. The convulsions have just ceased. Pulse 140; respiration 66.

Death took place at 2.42, from cessation of the respiratory movements; the fæces were involuntarily discharged a few minutes previously. The thorax was immediately opened, and phenomena almost precisely similar to those observed in Experiment XVIII. were noted.

B.—Death caused by very large doses.

Experiment XX.—A dog, full-grown, weighing about 15 lbs., and in excellent health, had ʒss of chloroform injected into his right pleura. The animal gave one cry and then lay still for a few moments; but in less than forty seconds the eyeballs turned up and became fixed, and all the limbs were seized with violent clonic convulsions. Respiration was gasping; the heart made, as it were, a short rapid *run* of feeble contractions, and came to a dead stop, less than seventy seconds from the moment of injection. The thorax was quickly opened, and the last quiver of the auricles was witnessed. No amount of irritation, by pricking or by dripping cold water on the surface of the heart, nor by pricking or roughly rubbing the thoracic ganglia of the sympathetic, nor by stirring up the cardiac plexuses with the handle of a scalpel, was able to reinduce the smallest visible contractions. The phrenic nerves were totally inirritable. Although the abdomen was quickly opened, no peristaltic movements of the bowels could be seen.

Experiment XXI.—A terrier, of about the same size as that used in Experiment XVIII., and quite healthy, was placed in a very large glass jar, charged with seven per cent. of chloroform vapor. The animal gave one or two convulsive starts, and fell. It was removed to the outer air in less than a minute from its introduction to the jar; its limbs were stretched out in a state of tonic spasm, the head was bent on the chest, and the whole body rigid. The heart was beating in a mere flutter, and stopped in a few seconds. On opening the chest, it was found relaxed and motionless; no irritation would re-excite contractions; not even strong irritation of the sympathetic ganglia. Immediately after death, the body became entirely relaxed: no note is preserved of the time when *rigor mortis* commenced.

Experiment XXII.—A sparrow (the subject previously of Experiment I., but which had quite recovered) was plunged into a jar charged with 4·5 per cent. of chloroform vapor. It made not a single struggle, but fell to the bottom of the vessel, and being immediately removed was found perfectly dead. The chest was instantly opened, but the heart was found motionless and inirritable.

Experiment XXIII.—A rabbit, full-grown, was plunged into a jar, which had been charged with 8 per cent. of chloroform vapor. In forty-five seconds it was violently convulsed, and on being quickly removed to the outer air it was found to be completely dead. The heart on inspection was not only motionless, but inirritable by any ordinary means; electricity was not tried.

Experiment **XXIV.**—**A full-grown,** but rather small, rat **had twenty** minims of chloroform (Bell) injected into its peri**toneal** cavity. The animal gave a scream, and after a few instants' repose became tetanically rigid—the tail extended and stiff, the limbs stretched out, and the toes spread widely. The heart ceased beating within one minute from the injection of the fluid, and no further attempt at respiration **occurred.** On opening the chest the heart was found nearly inirritable.

Experiment XXV.—Thirty minims of chloroform (Bell) were injected into the pleural cavity of a full-grown guinea pig. The operation was carefully performed, and seemed to **give** no pain. Fifty seconds later a general muscular tremor occurred: a few seconds later the limbs were suddenly stretched out, quite rigid, and the neck drawn backward and fixed. At this moment, the finger laid on the chest-wall **detected** the heart merely fluttering; in a few moments it could no longer be felt. The chest was opened; the heart **was** found motionless and inirritable.

2. *Fatal Narcosis in Man.*

The observations which I have to offer on this subject consist simply of an analysis of the facts, few, but, as I think, significant, which bear on the matter, and which have been noted by myself. It does not lie within the scope of my present purpose to investigate the conflicting opinions which have been given by various eminent authors as to the cause of fatal accidents in surgical anæsthesia; and I feel the less tempted to undertake this task, because the Committee appointed by the Royal Medico-Chirurgical Society is about to publish a report, in which it will be accomplished far better than it could be by me. I can only speak of what I have seen, and my remarks will include—(A) The history of one fatal accident which I witnessed; and (B) An analysis of *all* the cases in which, in my own practice, really alarming symptoms occurred.

A. *Case.*—A girl, aged seventeen, was placed under the influence of chloroform (with Snow's inhaler), in King's College Hospital, in order to the cauterization of some mucous tubercles on the labia with nitric acid. One drachm was placed in the inhaler, and the patient was rather quickly got under its influence, though, as I was not the administrator, I cannot speak certainly to the time employed. The operation was rapidly performed, and nothing particular was noted in the patient's appearance except considerable pallor; but in truth, as the girl had twice successfully inhaled chloroform before, it is likely that rather less attention was paid to her state than would otherwise have been the case. The house-surgeon, the dressers, and myself remained talking in the ward for a minute or so, and during this time the patient moved her legs in a mechanical kind of way, and the urine was discharged somewhat convulsively. These occurrences prevented any apprehension of mishap, and we all left the ward. Almost immediately after we had got down stairs the nurse followed to ask for some medicine, and mentioned casually that the girl looked very pale. The dresser ran up at once, and found her quite pulseless, with blanched lips and fixed eyes, the pupils very widely dilated. Artificial respiration and galvanism were tried for a protracted period, but without producing the slightest sign of returning animation.

The post-mortem examination in this case discovered the existence of considerable thickening of the mitral valve of the heart; but as the girl had twice before taken chloroform without ill effect, I think this fact has little value, especially as in my own experience I have never found any evil results from administering chloroform to patients suffering from similar diseases. The dresser who gave the anæsthetic was a very careful and skillful man. I therefore agree with Dr. Snow (who has also reported this case) in the opinion that the entire blame should be thrown upon the condition of the inhaler, which, on subsequent examination, was found to have got out of order in such a way that the patient might readily have breathed a highly concentrated atmosphere of chloroform vapor.

Dr. Snow, in his comments on this case, raises the question whether death took place *early*, or whether there was really, as there seemed to be, a movement toward recovery followed by a fatal relapse. He inclines to the former opinion, and I now agree with him entirely. At the time I was inclined to take the movements, &c. noticed after the conclusion of the operation, as distinct evidence against this view, but subsequent experience in the rapidly fatal chloroformization of animals has assured me that similar phenomena often occur after partial or even total arrest of the heart's action. I believe that the heart was suddenly overwhelmed by the chloroform at the time when anæsthesia was pronounced complete enough for the performance of the operation, and that it never rallied, although unfortunately no evidence could be procured as to the state of the pulse at this moment.

B. The number of cases in which, under my own hands, alarming symptoms occurred in the course of chloroform-narcosis was 21, out of a total number of 3,058 administrations. The following table exhibits at a glance the whole series :—

	Subject.	Nature of Operation.	Period of Occurrence of Symptoms.	Nature of Symptoms.	Duration of Symptoms.	Mode of Administration.
1	Child, æt. 2 years.	Circumcision.	3 minutes from commencement of inhalation.	Weak, fluttering pulse; gasping respiration; great pallor. Sudden.	15 minutes.	ʒi on a handkerchief.
2	Child, æt. 6 months.	Harelip.	2 or 3 minutes from commencement of inhalation.	Flickering, unrhythmical pulse; gasping breathing; livid lips. Sudden	10 minutes.	ʒi on a sponge.
3	Man, æt. 40 years.	Hæmorrhoids.	End of operation; on a second application after partial recovery.	Sudden arrest of pulse and breathing; livid lips; jaw dropped.	25 minutes.	Snow's inhaler, ʒiij and ʒiss. Artificial respiration.)
4	Lady, æt.52 years.	Amputation for cancer of breast.	Middle of the operation; but little blood lost.	Sudden intermittence of pulse; great pallor.	5 minutes.	ʒi on a handkerchief.
5	Lady, æt.32 years.	Varicose aneurism of lip and cheek.	Less than half a minute from the commencement of inhalation.	Sudden intermittence of pulse; pallor; labored respiration.	10 minutes.	ʒi on a sponge.
6	Man, æt. 60 years.	Hæmorrhoids.	Near the end of the operation.	Sudden pallor; flickering, running pulse.	7 or 8 minutes.	ʒiij in Snow's inhaler, then ʒss on lint.
7	Girl, æt. ?	Removal of lower jaw.	3½ minutes after commencement. 1 minute after induction of full anæsthesia.	Sudden flickering of pulse; slow and gasping respiration; great pallor.	5 minutes.	ʒij on lint. (Rapid anæsthesia.)
8	Girl, æt. 19 years.	Plastic operation on face.	After a very few deep inhalations.	Sudden pallor; livid protruded tongue; slow, gasping respiration.	5 or 6 minutes.	ʒij on lint.
9	Man, æt. 38 years.	Radical cure of hernia.	Middle of the operation.	Sudden pallor; very hurried, weak pulse (no large hæmorrhage).	A few minutes (very sick for some hours).	ʒij in Snow's inhaler, afterward ʒss on lint.
10	Man, æt. 40 years.	Lithotomy.	2 minutes from the beginning of inhalation.	Irregular, weak pulse; deadly pallor.	12 minutes.	ʒi on lint.
11	Man, æt. 22 years	Amputation of forearm.	After a few inspirations.	Sudden insensibility; pallor; failure of pulse.	10 minutes.	ʒi on lint.

	Subject.	Nature of Operation.	Period of Occurrence of Symptoms.	Nature of Symptoms.	Duration of Symptoms.	Mode of Administration.
12	Man, æt. 43 years.	Fatty tumor removed from shoulder.	Inhalation continued for 6 minutes.	Extreme dilatation of pupils; stertorous, gasping breathing; pulse pretty regular.	4 minutes.	ʒij in Snow's inhaler.
13	Man, æt. 31 years.	Circumcision.	Middle of operation.	Sudden pallor and failure of pulse, then of respiration.	6 minutes.	ʒi on lint, two additional doses of ʒss each.
14	Woman, æt. 29 years.	Removal of scirrhous breast.	After a few inspirations.	Extreme pallor; intermittent pulse.	5 minutes.	ʒi on lint.
15	Girl, æt. 22 years.	Necrosis.	After a few inspirations.	Faintness; hurried respiration.	3 minutes.	ʒi on lint.
16	Boy, æt. 15 years.	Lithotomy.	1 minute from formation of complete anæsthesia.	Sudden pallor and failure of pulse.	5 minutes.	Lint. ʒi first, then ʒss additional.
17	Woman, æt. 32 years.	Plastic operation on perineum.	1½ minutes from commencement of inhalation.	Simultaneous failure of pulse and breathing; jaw fallen.	20 minutes, (Artificial respiration.)	ʒi on lint.
18	Man, æt. 56 years.	Perineal section.	Just after formation of anæsthesia.	Sudden pallor; failure of pulse; livid lips; gasping respiration.	15 minutes.	Lint. ʒi and ʒss additional.
19	Woman, æt. 42 years.	Removal of scirrhous breast.	Sudden and violent epileptiform convulsion at end of 1st minute of inhalation.	Pallor; pulse very weak after fit.	8 minutes.	Weiss's inhaler (Out of order.)
20	Woman, æt. 37 years.	Removal of breast.	2 minutes from commencement of inhalation.	Sudden pallor; failure of pulse; respiration gasping.	3 or 4 minutes.	ʒi on lint.
21	Man, æt. 21 years.	Evulsion of toe-nail.	Middle of operation.	Sudden pallor; gasping respiration; fluttering pulse.	20 minutes.	ʒij and ʒss on lint. (Artificial respiration.)

Of these twenty-one patients, whose lives were placed more or less in jeopardy, it will be observed that only five took the anæsthetic through an apparatus of any kind. In four of these cases Snow's inhaler was used; in the remaining one Weiss's instrument, which is a modification of Snow's, was employed. When I add to this the observation that out of the total of 3,058 administrations, Snow's inhaler was employed above 1,800 times, and Weiss's about 400 times, it will be seen that this fact has a certain significance. But this significance is much increased on an analysis of the five cases in which alarming symptoms occurred during the use of an inhaler. In Case 3 (*vide* table) it was discovered that the inspiratory valve had probably become fixed for a sufficient time to allow of a considerable accumulation of chloroform vapor in the tube leading to the face-piece: as a result of this the patient struggled ineffectually for some moments to obtain any air at all, and then, removing the obstacle by a forcible inspiration, received an atmosphere very highly charged with the vapor. The symptoms followed instantaneously. (This was the most serious case that ever occurred under my care; the patient was only saved by the most vigorous use of artificial respiration, by Marshall Hall's process.) In Case 6 the inhaler had been safely used for the induction of anæsthesia; but the patient showing signs of an inconvenient return of consciousness, and the tube of the instrument unluckily becoming detached, a small additional quantity (ʒss) was administered on lint, and the dangerous effects on the pulse almost immediately supervened. In Case 9 precisely the same order of events occurred. Case 12 was one in which the inhalation proceeded regularly enough, but was inadvertently carried too far, producing serious depression of the respiration, without any noticeable irregularity in the heart's action. Here there was imminent danger of death *from apnœa*, such as will be described in the third series of experiments on animals. In Case 19, Weiss's inhaler was employed; but owing to a defective arrangement of the blottıng-paper in the interior, liquid chloroform overflowed, and a small quantity was probably drawn into the larynx. In all the remaining cases in the table, the chloroform was given either on lint, a handkerchief, or a sponge—that is to say, in a manner which makes no adequate provision for *the regulation of the strength of the vapor inhaled.*

CONCLUSIONS.

1. The narcotic action of chloroform when studied upon the large scale, and in different animals, presents uniform phenomena; or rather, the variations observed are merely such as may be assigned to simple causes.

2. The general course of chloroform-narcosis is similar to that of the narcosis induced by æther; it consists of the regular and progressive extinction of the vital properties of the various portions of the nervous system in the order described in the remarks on that agent.

3. Two distinct lines of narcotic progression may, however, be traced in the action of chloroform upon the animal organism, the development of the one or the other depending entirely upon the rapidity with which the arterial circulation becomes charged with the narcotic agent.

A. When the impregnation of the blood takes place with *moderate* rapidity, the sympathetic nervous system is the *ultimum moriens,* and death *begins at the lungs.*

B. When, on the contrary, the circulation becomes very rapidly charged with a large proportion of chloroform, the narcotic effect may fall with such force upon the sympathetic nerves as to extinguish their vitality at once.

4. The greatest possible importance attaches to this distinction; for one of the consequences of the *latter* occurrence is the production of instantaneous paralysis of the heart. This catastrophe may be induced, at pleasure, by the inhalation of an atmosphere highly charged with the vapor, or the injection of a large dose of liquid chloroform into the peritoneum or pleura.

5. That this action is exerted locally upon the sympathetic nerves in the heart is rendered probable by the phenomena (which appear to be parallel to this*) of chloroform-diabetes. As in the case of the production of diabetes by æther, this latter effect is more decidedly manifested when the chloroform is injected into the portal vein than when it is inhaled.

6. From the nature of the symptoms which have presented themselves in all the cases in my own experience where life was manifestly threatened, from the phenomena observed in the only fatal case witnessed by me, and from the fact that in the immense majority of reported fatal cases the first symptom of danger was confessedly the failure of the pulse and the blanching of the countenance, the conclusion appears strongly indicated, that paralysis of the heart is *the* source of danger in surgical chloroform-narcosis.

7. If this be the case, there can be little doubt in the mind of any one who has carefully perused the observations recorded in the preceding pages, that *the* desideratum for the perfectly safe administration of chloroform in surgical practice is an apparatus which will supply an atmosphere† of uniform and moderate strength, and not above 3·5 per cent. for inhalation. The practice of intrusting the induction of anæsthesia to unskilled persons *without* the protection of such an apparatus is to be censured in the strongest terms.

In addition to the above formal conclusions, which are the logical result of experiments which have actually been detailed, I desire to express one or two opinions based upon evidence which the necessary limits of this work forbid me to insert.

8. It is, I believe, quite useless to expect that good will result, in cases of grave danger from chloroform, attended

---

* The arguments for this parallelism are the same as those which have been used in relation to æther. *Vide supra*, p. 277, et seq.

† The apparatus invented by Mr. Clover is by far the nearest approach yet made to theoretical perfection, and it also gives decidedly the best practical results that have been obtained by any mode.

with sudden and complete, or nearly complete, failure of the pulse, from any measure short of the immediate and energetic performance of artificial respiration.

9. Even this measure will probably be useless in cases where a very profound paralyzing effect has been produced on the heart. I have repeatedly recalled animals to life by means of artificial respiration, where there was apparent death by apnea, and the heart's action had only *gradually* declined. I have *never once* succeeded in resuscitating an animal whose circulation had *suddenly* come to a standstill simultaneously with, or prior to, the cessation of breathing.

10. It is my firm persuasion that, with proper care, chloroform may be safely administered *to any patient who is fit to undergo an operation at all, whether there be any existing disease of heart, lungs, or brain, or not.* I have never allowed the known existence of such disease to prevent my administering it, and I have never found any evil result. I entirely concur with Dr. Snow's opinion on this subject.

PART III.

The stimulant Action of Chloroform.

I HAVE now to speak of the effects of chloroform given in doses too minute to produce any symptoms of narcosis. This is a subject which has been undeservedly neglected, owing to the natural prejudice which inclines us to regard every effect produced by this agent as necessarily involving some degree of anæsthesia. It is certain, however, that the phenomena which we are now to consider are produced quite independently of any such action.

These phenomena may be classed under three heads :—1. Relief of certain forms of pain. 2. Arrest of convulsive muscular movements. 3. Restoration of the natural movements of the uterus in parturition, when they are deficient.

1. *Relief of certain forms of Pain.*

It is a very common observation that chloroform, inhaled or applied to the unbroken skin, is a very effective remedy for neuralgic affections, and it is usually supposed that the benefit obtained is due to its narcotic action. Careful observation, however, leads me to a different conclusion. I have always noticed that when chloroform-inhalation, for instance, relieves a severe tic, it does this in one of two ways—either it removes the pain during the very earliest inspirations, and before the slightest degree of true anæsthesia is produced, or it fails altogether to produce any impression till unconsciousness arrives. The former is the true medicinal action, and it is certainly a form of stimulation, since the amount of

chloroform absorbed into the blood in such cases is quite
insufficient to produce any paralyzing influence upon the sensory nerves.   It is exactly paralleled by the effect of an application of chloroform, diluted with seven or eight parts of
some saponaceous or greasy liniment, to the unbroken skin.
In the latter instance, a very small quantity of chloroform
rapidly permeates the skin, and acts as a stimulant to the
capillary circulation : in fact, the effect, as nearly as possible,
resembles that of a mustard-plaster, except in its rapidity.
I can speak from personal experience to the fact that this is
the result produced, and that not the slightest numbness is
occasioned when the remedy acts most favorably.

But the most remarkable instance of the anodyne influence
of small doses of chloroform is seen in the relief which it gives
to the sufferings of parturient women.   A great deal of unnecessary prejudice against its administration in labor has
been occasioned by the mistaken notion that it is necessary
to stupefy the patient, and to put her life more or less in jeopardy.   This is not the case.   It is only necessary at the commencement of each pain to charge a Snow's inhaler with fifteen
minims of chloroform, and allow the patient to inhale : this
arrangement permits only an extremely small quantity of the
vapor, plentifully diluted with air, to enter the lungs : not the
slightest unconsciousness or disturbance of intellect need be
produced.   I shall recur to the subject of chloroform in parturition presently.

2. *Arrest of convulsive Muscular Movements.*

I have already alluded, in the chapter which describes the
general phenomena of stimulation, to the efficacy of chloroform in arresting epileptic convulsions, and I have explained
that this effect is attainable by doses which are non-narcotic,
for the patient actually recovers consciousness under the action
of the remedy.   This is but one example, for I believe that
convulsion of every kind, and all the varieties of spasm, are

capable of being temporarily arrested in the same way. The following experiments are interesting in relation to this question :

Experiment XXVI.—Preliminary trials had proved that doses of $\frac{1}{10000}$th of a grain of strychnia injected into the bodies of frogs will cause tetanus, on the average, in two minutes. A glass jar was charged with an atmosphere of 0·5 per cent. chloroform vapor, and the above-mentioned quantity of strychnia being injected into a frog's belly, the animal was placed in the jar. No symptoms of tetanus occurred till the animal had been nearly ten minutes in the jar; by this time anæsthesia was almost complete. The symptoms of strychnia poisoning now began to be developed; the animal was, however, only affected with frequently recurring *tremors*, slight jerking movements of the head, and extension of the limbs during the attacks of tremor. Death took place by apnea, apparently from the influence of the chloroform, but rigor mortis almost instantly followed the arrest of the heart's action.

Experiment XXVII.—A glass jar was charged with one per cent. chloroform vapor, and a frog, as nearly as possible of the same size as the one employed in the last experiment, and poisoned with the same dose of strychnia, was introduced into it. The animal became narcotized in the usual way, and was apparently quite insensible in eight minutes. In four minutes longer no symptoms of strychnia poisoning had supervened, but immediately after this slight tremors were noticed, and the limbs were extended. Respiration came to an end in eighteen minutes from the commencement of the experiment; the animal was quickly removed from the jar and the chest was opened; the heart ceased acting almost immediately afterward. Irritation of the motor nerves (*e.g.* sciatic) caused no contraction of the voluntary muscles, and irritation of the sympathetic produced no effect on the heart; all the muscular structures passed rapidly into the condition of rigor mortis.

Experiment XXVIII.—A similar dose of strychnia was injected into the cavity of a frog's body, and *one drop** of chloroform was injected (suspended in five minims of thin mucilage) three minutes later, just as the first decided tetanic spasms were developed. The result was a complete cessation of all convulsive movements for nine or ten minutes, after which the limbs became rigidly extended, and a series of tetanic spasms occurred which proved fatal in four minutes and a half. This experiment was repeated on a second frog with similar results.

From the above experiments, and from others which were made with an aqueous solution of opium instead of strychnia, I derive conclusions similar to those which resulted from the parallel experiments with æther, viz.—(1) Chloroform has no **direct antidotal action to strychnia.** (2) In large doses it may indirectly prolong life by inducing paralysis rather than convulsion: this is owing to its causing a more complete destruction of the vital conditions of motor nerves, as already explained. (3) **In small** stimulant **doses** chloroform has the **power,** temporarily, to arrest the convulsions of strychnia, *without* inflicting **damage on the vitality of the nerves: its** action in this case is therefore, *pro tanto,* beneficial.†

3. *The influence of small doses of Chloroform in favoring the progress of Parturition.*

Used in the way above described, I am satisfied, from very considerable experience, that it materially increases the force and regularity of the uterine contractions, and that its action by no means only or chiefly consists in the relaxation of the

---

* Between ¼ and ½ of a minim.

† Very lately I have had the opportunity of repeating these experiments on mammalia (rats, cats, and rabbits) with results completely similar to those obtained on frogs. The action of narcotic doses of chloroform only renders the action of strychnia more certainly fatal, though somewhat slower: in some cases clonic convulsion is substituted for tetanus. The action of a single stimulant dose will temporarily arrest or diminish convulsion, without doing any harm.

external passages. **Again and** again I have seen the contractions of the uterus, which had **been weak** and irregular, become strong and effective, at the same time that their painfulness was greatly diminished or removed, under the influence of minute doses of chloroform. On the other hand, after careful observation, **I feel** confident **that** injurious **results** have often followed the injudicious perseverance of an administrator who could not be content without producing unconsciousness by the use of narcotic doses. In such circumstances, **the uterus is apt either to become** inert or to act in a hurried, weak, and inefficient manner.* **The only case in** which the narcotic action of chloroform should be induced is where *operative* interference is required, and where, consequently, absolute **stillness on the part of the patient is necessary.**

* This principle was partially developed by Dr. Murphy in his pamphlet, "Chloroform in Practice of Midwifery." (London: Taylor and Walton. 1848.) It is also the basis (though this is not clearly explained by the author) of the success of Dr. Townley's method, described in "Parturition without Pain." (London: John W. Davies, 1862.) But there is a good deal of unnecessary mystification in the very complex directions for preparing the "anodyne mixture."

ALCOHOL.

RESEARCHES ON THE PHYSIOLOGICAL ACTION OF ALCOHOL.

THE inquiries which we have made into the *modus ope-randi* of æther and chloroform are a useful preparation for the study of the effects of alcohol. The latter substance, in large doses, is an *anæsthetic* equally with the two former— · that is to say, a narcotic which operates with especial force upon the sensory nervous tract ; but as its poisonous operation is less frequently the subject of scientific examination than that of æther or chloroform, the resemblance is less familiar to us than it would otherwise be. Moreover, the striking nature of the effects upon consciousness which alcohol produces are calculated to divert attention from its anæsthetic operation.

The first part of the following researches is therefore directed to the object of placing the narcotic action of alcohol in a clear light, and tracing the limits of this action. The second part is devoted to the consideration of its non-narcotic effects.

PART I.

Narcotic Effects of Alcohol.

1. *Induction of Narcosis.*

A.—Alcohol administered by the Stomach.

EXPERIMENT I.—A dog, full-grown and healthy, weighing 10 lb. 4 oz., had 6 oz. of a mixture, consisting of equal parts rectified spirit of wine (P. L.) and water, introduced into his stomach by an œsophageal tube, at 1 P.M. No food had been taken for four hours previously.

1.4. The animal is obviously affected : he staggers in walking, looks puzzled, and frequently falls down. On examining him carefully, it is seen that the hind quarters are very weak, and, moreover, the skin of the hind limbs is partially insensitive. Respiration 24, circulation 140.

1.6. The dog lies extended on the floor, very drowsy, but capable of being roused ; the hind limbs are completely paralyzed, the fore limbs retain a slight degree of voluntary power. The conjunctiva is fully sensitive still, but the skin about the mouth and face seems to be entirely paralyzed as to sensation. The tongue is protruded, and the dog slavers somewhat.

1.7.30. The animal falls on its side, comatose and snoring. The conjunctiva is now equally insensitive with other parts. Respiration 20; circulation 184, tolerably strong,

While the animal remained in this condition, it was determined to examine whether any part of the body retained sensibility. It was found that the whole ano-genital region was still so far sensitive that a slight whine of pain was elicited by pinching with forceps in any part of it.

The pupil, during the first ten minutes after the effects of the alcohol became apparent, was strongly contracted: at the end of this time it began to dilate, and at 1.25 it was seen to be perfectly dilated, and very little sensitive to light. Respiration had now risen again in frequency (to 30); heart's action 200, and somewhat irregular.

1.32. The dog made a slight struggle and vomited what was evidently a small part of the alcohol; perhaps there was half an ounce of fluid, part of which was mucus. The alcoholic smell was distinct. There was a slight improvement for a few minutes after this, partial consciousness apparently returning: the anæsthesia of the surface, however, remained complete. Even the conjunctiva was perfectly insensitive.

From this moment nothing worthy of note occurred, except that the hinder limbs were affected with a continuous tremor for a short time. Respiration gradually declined in frequency and became gasping; it finally ceased at 3.5, two hours and five minutes from the administration of the alcohol. The heart was then beating 64 per minute, and continued to act slightly for a few minutes. It remained irritable for some moments later.

Experiment II.—Twenty minims of Sp. vin. rect. (P. L.) were injected, drop by drop, into the stomach of a healthy full-grown rat, at 12.5 P.M. The animal ran about nimbly for about ten minutes, it then became still and drowsy.

12.33. Respiration hurried, heart's action rapid, eyes dim, consciousness impaired. When made to walk, the animal staggers. The posterior limbs are very weak.

12.34. The hind limbs are palsied as to motion and sensation.

12.36. Complete insensibility of the whole surface of the body; motor palsy of all the limbs. The animal was still, however, partly conscious. Breathing and circulation extremely rapid.

12.38. Violent clonic convulsions of the limbs; respiration alternately suspended and very hurried; heart's action very rapid, feeble, and irregular.

12.42. Respiration ceased. Chest immediately opened. Heart beating 80 per minute, with some force.

12.47. Only the auricles contract, feebly.

12.48. Contraction ceased; heart remained irritable some **minutes longer.**

Experiment III.—Five minims of Sp. **vin. rect.,** diluted with an equal quantity of water, were dropped, gradually, into the pharynx of a healthy white mouse, of average size, at 2.40 P.M. The animal got on its feet and ran about immediately after the administration, and for two minutes no symptoms appeared. After this the mouse became drowsy, and lay down.

2.42. Skin of the muzzle perfectly insensitive.

2.43. The animal was plainly paralyzed in its hind limbs; on stirring it up, it made feeble efforts with the fore legs. Skin everywhere insensitive. The conjunctiva of the eyeball still retained its sensibility.

2.45. The animal was seized with violent clonic convulsions.

2.48. **The convulsions had now ceased;** the animal lay on its side profoundly unconscious and completely insensitive; respiration and circulation extremely hurried.

2.50. Respiration gasping **and** extremely difficult, 14 per **minute;** heart's action rapid and unrhythmical.

2.51. Respiration ceased. The thorax was quickly opened, **and the heart was seen still beating:** it continued to act for **three minutes** longer.

Experiment IV.—A healthy full-grown rabbit had one **ounce of the Sp. vin. rect. (P. L.) injected slowly into its stomach at 3.35 P.M.**

No vomiting took place; the animal seemed lively and undisturbed for about two minutes; it then became excited and **tried to run away, but** staggered and **fell** repeatedly.

At 3.54 the hind legs gave way and slid from under the rabbit: they were found completely paralyzed as to sense and motion. Respiration very hurried; circulation so rapid as to be uncountable; consciousness still perfect.

3.59. The animal lies on its side partly comatose; when roused it scrambles feebly with its fore legs. Surface of the eyeball still fully sensitive; face profoundly insensitive. Sensibility of the ano-genital region apparently not at all less than in health. Respiration 60, regular; heart's action 170, and rather weak.

4.30. The rabbit had continued profoundly comatose since the last report. Pinching the skin near the anus still elicits slight signs of pain.

Experiment V.—Ten drops of Sp. vin. rect. were injected slowly into the throat of a healthy sparrow, the operation being made in four stages two minutes apart from each other. At the first dose of two drops the bird was almost instantaneously affected with loss of locomotive co-ordination; it staggered and could not fly properly. Three drops more almost immediately produced coma, with slight clonic convulsions. By the time the whole had been injected respiration was effected in a series of gasps, the heart beating very rapidly. The comatose condition lasted for more than half an hour, breathing gradually becoming somewhat stronger; at the end of this time the bird began to move, and in a few minutes got on its feet, though staggering very much. Ten minims more alcohol were now injected at once (though slowly) into the throat. Respiration immediately became very rapid and fluttering, then gasping, and in a minute and a half from the second dose it ceased. The heart continued to beat for about a minute longer. Insensibility of the surface was complete from the moment of the first occurrence of coma till the time of death.

Experiment VI. was the one already detailed,* which I made upon myself with ʒiss of whisky taken upon an empty

* Vide page 187.

stomach in the morning. In this instance I used a quantity of alcohol so small as I should not beforehand have supposed capable of producing the poisonous results, had it not been for the report of Dr. Edward Smith's experiments, which made me chary of venturing on a larger dose under such conditions. The poisonous effects were fully developed, though not very lasting; and it became obvious to me from this, and from a repetition of the experiment, that the time of day at which the experiment was made (as well as the *entire emptiness of the stomach*) caused the system to be unusually sensitive.

From the account already given of the symptoms observed in this experiment, it will be obvious that they were essentially the same with those observed in the animals above spoken of. The same trouble of the brain, the same spinal paralysis proceeding from below upward, was noted. The principal phenomena which occur in man and are not readily recognized in animals are the partial paralysis of the trigeminal and hypoglossal nerves, and also of the cranial sympathetic, which are among the *early* phenomena in the human subject. With regard to the trigeminal, however, there was considerable evidence in several of the experiments that it was probably affected as in man. (*Vide* Experiments I., III., and IV.)

In the following experiment, partial paralysis of this nerve was clearly marked as an early symptom.

Experiment VII.—A remarkably large, strong, and active cat had two ounces of a mixture of equal parts of Sp. vin. rect. and water injected through an œsophageal tube into its stomach, at 12.30 P.M. The animal had been fasting for some hours.

12.10. The cat had been quite lively up to this time, but it now appeared drowsy, and its head drooped forward, the muzzle resting on the stone floor of the room in an awkward position. On pinching the skin round the mouth, or pulling the whiskers, no sign of feeling was evoked; the animal was raised on its feet, when it became evident that the posterior limbs were nearly paralyzed. The anterior were little affected. Complete paralysis of motion and anæsthesia of the whole surface (but not of the conjunctiva) was produced by 12.15. Two minutes later the conjunctiva was also insensitive; the pupil had been widely dilated from the first occurrence of motor paralysis, and was now found insensitive to light.

Death took place by cessation of the respiratory movements (death by apnea) at 4.45.

B.—*Injections into the Peritoneal Cavity.*

Experiment VIII.—Ten minims of alcohol (Sp. vin. rect. P. L.) were injected into the peritoneal cavity of a healthy white mouse, with a Wood's subcutaneous injection syringe, at 2.48. The animal ran away nimbly when released, and continued running round the table for nearly a minute; it then shivered and stood still, and its hind limbs slipped from under it, paralyzed. Thirty seconds later the mouse fell on its side, and all its limbs were affected with clonic convulsions: respirations very hurried; heart's action not to be counted. At 2.50 the respirations ceased suddenly: the chest was instantly slit open and the heart was seen to pulsate feebly for two or three minutes.

Experiment IX.—Five minims of the Sp. vin. rect. were injected into the peritoneal cavity of a healthy white mouse at 2.56 P.M.

2.58. The animal was comatose, paralyzed, and affected with clonic convulsions; respiration very hurried.

2.59. Respiration ceased suddenly. Heart continued to act feebly for two or three minutes; it was slightly irritable after cessation of its action.

Experiment X.—Two minims of the Sp. vin. rect. were injected into the peritoneal cavity of a healthy white mouse at 3.2 P.M.

3.6. The animal has lain down and is almost completely paralyzed, and partly comatose, but can be roused. Respiration hurried ; circulation very rapid.

3.18. Respiration, which had been gasping and slow for some minutes, ceased ; the heart continued to beat for nearly ten minutes longer.

Experiment XI.—Thirty minims of Sp. vin. rect. were injected into the peritoneal cavity of a full-grown healthy rat at 12.8 P.M.

12.9 The animal has fallen on its side, breathing very rapidly; when touched it struggles rather feebly, with its fore limbs only. Muzzle nearly insensitive.

12.12. Surface of the body everywhere completely insensitive ; limbs relaxed ; respiration and circulation very rapid.

12.21. The rat was seized with clonic convulsions of all the limbs, which lasted intermittingly for five minutes. The limbs then suddenly relaxed ; respiration became gasping, and in a minute or two ceased. The heart beat for six minutes longer.

Experiment XII.—Twenty minims of Sp. vin. rect. were injected into the peritoneal cavity of a full-grown rat at 2.3 P.M.

2.5. Breathing very hurried, heart's action also very rapid. The rat was very uneasy.

2.6. The animal lay on its side, breathing very rapidly, pulse uncountable, skin everywhere insensitive.

3.0. The animal lies on its back. Respiration very rapid and shallow, almost exclusively abdominal; there is still partial consciousness.

3.50. The general condition had remained much the same. There were now slight convulsive twitches of the muscles of the face.

4.45. The general condition had been getting steadily though slowly worse ; respiration and circulation more hurried and feeble. At this moment slight general clonic convulsions occurred, and were repeated several times during the next few minutes.

5.0. The occurrence of the convulsions seems to have marked the point of greatest depression, for the general condition had since improved considerably. Respiration 70; pulse 130 and of pretty good force. Consciousness had returned to a considerable extent, and the animal made voluntary movements with its forelegs. Recovery went on without intermission from this time, and was complete by 8 P.M.

Experiment XIII.—A full-grown, large, and active cat had 5i of Sp. vin. rect. injected into its peritoneal cavity. No symptoms whatever occurred, except that the cat vomited once or twice, and seemed rather drowsy for a few hours; and at the end of three days the animal was active and healthy as ever, eating its food readily.

Experiment XIV.—The same cat had *one ounce* of Sp. vin. rect. injected into its peritoneal cavity, at 2.30 P.M. After the operation, the animal (which had not appeared hurt, but only frightened) ran round the room, jumped on a shelf, and took refuge behind some jars.

2.34. The animal appearing drowsy, was lifted on to the floor, when its posterior limbs were found paralyzed as to motion and sensation, and its muzzle quite insensitive. The animal was fully conscious, but gave no sign of pain; when handled it struggled with its fore limbs to escape.

2.35. Complete paralysis of all the limbs; eyeball still sensitive, pupil widely dilated, but retaining some sensibility to light. Respiration 24, quite regular. Heart's action 120, regular.

2.40. Complete anæsthesia and unconsciousness; eyeballs turned up, and rotating convulsively in alternate movements, outward and inward; pupils dilated and *fixed*, conjunctiva insensitive, eyelids still irritable.

From this time forward no symptoms worthy of record are known to have occurred, except that, at 8.30 P.M. the right foreleg was noticed to be convulsively contracted. Death took place at about 6.30 A.M., or sixteen hours from the time of injection. Rigor mortis appeared about six hours later.

When the symptoms of alcoholic poisoning had fully developed themselves it was determined to test the animal's breath, in order to establish the fact of elimination; and an apparatus was applied, by means of which the expired air was made to pass through a solution of bichromate of potash in sulphuric acid. Unfortunately (as was afterward discovered) a mistake had been committed in the hasty manufacture of the test liquid, and the results were negative, or at least indecisive, but the alcoholic smell was distinctly perceived.

Eight hours after death the belly was opened. It contained between eleven and twelve drachms of serum slightly colored with blood; there was a very small amount of lymph deposited, here and there, upon the intestines. The cavity of the chest presented no abnormal appearance; there was a moderate quantity of dark-colored blood in the heart and great vessels. Five drachms of the peritoneal fluid, diluted with an equal quantity of water, were placed in a small retort, heated by a spirit-lamp. The heat was raised so as to coagulate as suddenly as possible most of the albumen present, and the distillate allowed to pass over to an appropriate vessel; a Liebig's condenser was used. The first two drachms of distillate being of a muddy color, were returned to the retort; and the albumen being now nearly all solidified, the distillation proceeded quietly. About two and a half drachms of a limpid liquid, having still an animal odor, were obtained; this was tested with a carefully prepared solution of one part bichromate of potash in 300 parts of strong sulphuric acid, and a very decided emerald green color was produced. To the remaining two drachms of distillate a quantity of carefully dried carbonate of potash was added, with a view to abstract the water and allow any alcohol present to float free on the surface. But no distinct layer of alcohol could be observed, after thorough subsidence; hence

it was judged that the quantity of alcohol present must have
been very minute.* Since nearly half of the contents of the
peritoneal cavity had been operated on, it was concluded that
but an inconsiderable portion of the large dose of alcohol injected sixteen hours before death had escaped absorption.
The brain and spinal cord were carefully inspected: they presented no remarkable appearance except that their substance
was everywhere rather unusually pale. The animal had lain
on its left side from the moment of its becoming insensible
till the post-mortem examination, and as a consequence of
this there was some congestion of the veins of the pia mater
over the left hemisphere of the brain. (Owing to circumstances it was impossible to make a chemical examination of
the nervous substance or the blood, in this case. For the
general purposes of the experiment, moreover, there was no
particular necessity for this.)

Experiment XV.—A fine healthy spaniel, full grown, had
one drachm of Sp. vin. rect. injected into his peritoneal
cavity. The operation did not appear to give any pain. The
only symptom which was produced was vomiting, an hour or
two later. But the animal continued apparently in full appetite and health, and the vomiting only recurred twice.
The dog was carefully watched at intervals for forty-eight
hours, and then appeared as well as if nothing had been done
to him.

Experiment XVI.—The same dog had *one ounce* of Sp.
vin. rect. injected into his peritoneal cavity. As before, the
dog vomited three or four times in the course of the next
hour or two; *but he showed absolutely no other symptom of
mischief*, and kept his appetite and activity perfectly.

Experiment XVII.—The same dog, forty-eight hours later,
had two ounces of Sp. vin. rect. injected into his rectum, at
3.35. Notwithstanding great precautions, nearly an ounce

---

* Dr. A. Dupré, who kindly gave me his assistance in this experiment, assures me
that if no more than 5 per cent. of alcohol had been present in the distillate, a distinct
layer would have been formed by it in the upper part of the test-tube.

and a half of this was shortly voided again. Roughly speaking, half an ounce may be said to have been retained.

3.45. The dog exhibited signs of incipient drunkenness, in the loss of co-ordinative power in walking; he ran away, and fell down several times. He then attempted to sit down, but his hind quarters were now so paralyzed that they slid away, and he rolled over. Pupils contracted, muzzle insensitive. Breathing regular, 20; heart's action 100, regular.

4.30. The dog was still quite conscious, but rather more paralyzed in his hind quarters, was drowsy, and lay down a good deal.

The symptoms of alcoholic intoxication reached their height at about 6 o'clock, when the animal was perfectly paralyzed in his hind quarters, semi-comatose, and perfectly insensitive as to his muzzle. From this time he began to recover, and by half-past ten was quite well.

This same animal, a week later (being in perfect health), was killed almost instantaneously by the injection of ʒss of chloroform into the peritoneal cavity. On examining the abdomen no traces of inflammation were visible except around the points of puncture of the abdominal wall. There was not a drop of fluid in the abdominal cavity.

The above experiments establish the general similarity of the narcotic action of alcohol injected into the peritoneum with that which it exerts when received into the stomach or rectum. Whatever inferiority of energy it shows in the former case appears to be due to the fact of its escaping from the lungs, in part, before it has time to reach the arterial circulation, and thereby influence the nervous system, whence it may happen that a considerable dose (owing to the slow rate of its absorption from the peritoneum) may fail to saturate the blood sufficiently to produce an anæsthetic effect. Why the rate of absorption from the peritoneum should be so different in the case of the cat and the dog respectively (a difference which has shown itself in every comparative series of experiments which I have made), is a question which claims future attention, but which I cannot at present answer.

The symptoms of alcohol-narcosis, as displayed in the experiments which have now been described, are very similar to those produced by chloroform and æther. There is the same gradually advancing sensory and motor palsy, commencing with the posterior (lower) portion of the body, and blended and confused with the effects on consciousness produced by palsy of the brain. This palsy of the brain is responsible for all the so-called phenomena of "mental excitement," as explained already in Chapter V. of this work. We must now notice a series of effects which also find their parallel in the action of chloroform and æther, viz. the various forms of sympathetic paralysis:—(*a*) Flushing of the face, increased heat of the ears, injection of the conjunctiva. (*b*) Excessive rapidity, irregularity, or preternatural slowness of the circulation. (*c*) Appearance of sugar in the urine from disordered liver function. (*d*) Changes in the pupil. (*e*) Excessive diuresis.

(*a*) Flushing of the face is an almost constant phenomenon in intoxication; and it is certainly one of the earliest, if not the very earliest recognizable symptom of that state in the human subject. In Experiment VI. this symptom was strongly marked, the face and the ears becoming visibly red and sensibly hot, at the same time perspiration broke out upon the forehead (and, later, upon the cheeks). The following experiment, being made with alcohol itself, is more decisive:

Experiment XVIII.—One ounce and a half of Sp. vin. rect., diluted with an equal quantity of water, was taken, upon an empty stomach, at 8.0 A.M. In rather less than ten minutes a sense of throbbing in the facial vessels was very distinctly perceived, and appeared to radiate over the whole face, which became overspread with a crimson flush; the ears also became very red; the temporal arteries throbbed to an unpleasant degree, and very shortly a visible perspiration

appeared on the forehead. The conjunctiva was strongly
injected, and the eyes were suffused with tears. (These
symptoms ushered in a considerable and very unpleasant
confusion of consciousness.) The congestion of the face persisted for more than half an hour.

There can be no doubt, I think, that the above phenomena
are due to paralysis of the cervical and sympathetic, and
they entirely agree with the symptoms observed in animals
under similar circumstances. Upon the average of twenty
experiments made upon cats, dogs, rabbits, and guinea-pigs,
I found that the thermometer placed within the ear marked
a rise of 3° 56′ Fahr., during the early stages of alcoholic
poisoning.

Dr. Edward Smith lays down the rule, as the result of his
extensive observations, that the action of alcohol is attended,
in all but a minority of cases, with dryness of the skin and
of the mucous membrane of the mouth. I cannot help de-
murring to this statement. In the earlier stages of alcoholic
narcotism I have far more frequently observed a sensible
perspiration on the brow, and usually on the cheeks; and
although it is perfectly true that with some persons the
slightest excess, far short even of intoxication, at once induces
a dry and harsh state of the skin and of the mucous membrane of the mouth, more commonly this effect is not produced till the stage of recovery from intoxication: the
proverbial parched mouth and dry skin of the drunkard is
an after-consequence of the debauch.

(b) The narcotic action of alcohol upon the heart follows
precisely the same rules as those which mark the influence
of chloroform and æther upon that organ. A narcotic dose
of alcohol always increases the frequency of the pulsations to
a decidedly abnormal extent (unless it be so extremely large
as at once to produce a profound degree of cardiac paralysis):
the effect produced appears closely analogous to that of rough
handling of the thoracic ganglia and the cardiac plexuses
after death by apnea, while the sympathetic nerves still

retain their "irritability." The kind of effect produced on
the heart varies, however, extremely—according as the circulation is slowly or rapidly charged with a large dose of
alcohol. In a large number of cases in the human subject
the excessive rapidity yields, long before death or the commencement of recovery, to a depression both in force and
frequency, which places it below the line of health. In the
rarer instances of an enormous dose rapidly absorbed, life
may come to an end without any other effect being produced
on the heart than that of a shock-like depression, which
brings it to a stand-still within a few moments or minutes.
In other cases, a rapid but feeble action of the heart continues till within a short period of the fatal termination, and
is suddenly succeeded by intermittence, and then cessation,
of contractions. It is to be noted, that excessive increase of
rapidity of the circulation is a far more prominent symptom
in the lower mammalia than in man.

(c) The production by the liver of a quantity of sugar
sufficient to produce diabetes is a phenomenon observed in
poisoning with alcohol, though not so commonly as in poisoning with either æther or chloroform. It is produced readily
enough, as Dr. Harley first observed, by the injection of
alcohol (in considerable concentration) into the portal vein.
It is by no means easy, however, to produce it by the administration of alcohol by the mouth, for this procedure does not
necessarily involve its absorption in anything like a concentrated form into the portal circulation.

I have paid considerable attention to this subject. After
repeating and verifying the excellent observations of Dr.
Harley, on the effect of direct injection of alcohol into the
portal vein of animals, I attempted to imitate the process on
my own person, by taking doses of from two to four ounces
of Sp. vin. rect. at 9 P.M., and carefully testing for sugar
three hours later, and also at 9 A.M. and 9 P.M. on the following day. Notwithstanding the decided narcotic effects, and
the great disturbance of the digestive organs which were

produced, there was not in any of these experiments (four in number) the slightest appearance of sugar in the urine, although this might easily have happened from the mere perturbation of digestion.

I have been able, however, on two occasions to detect the sugar in the urine of alcoholized persons. In two cases of very severe alcohol-poisoning which came under my own notice, urine removed from the bladder during the period of coma was found to be strongly saccharine. But I have repeatedly failed to find sugar in the urine of persons who were even so far intoxicated as to be unable to take care of themselves.

In animals there is a very considerable difference as to the facility with which diabetes can be produced by the introduction of alcohol into the interior of the alimentary canal; it is not difficult, however, to obtain this effect in the rabbit. Direct injection of anything like a large dose (*e.g.* ʒii Sp. vin. rect., diluted with equal parts, or even with three parts of water) *into the portal vein* of a dog or cat, infallibly produces diabetes, so far as my own experience goes, if the animal survive the operation for three or four hours; although in the case of the dog no visible symptoms of intoxication, or but slight ones, may present themselves.

I have also found sugar in the urine of a dog into whose peritoneal cavity ʒss of Sp. vin. rect. had been injected three hours previously.

These various considerations induce me to explain the diabetes produced by alcohol, in the same way as that caused by æther and alcohol. I believe that a strong local effect on the sympathetic nerves in the liver itself, or else a very profound degree of poisoning of the whole nervous system, is required to produce this effect; and in either case I cannot doubt that the influential cause is a *paralyzing* agent.

(*d*) The changes in the pupil produced by alcohol appear to follow very much the same course, on the whole, as those observed in anæsthesia produced by chloroform or by æther.

The final stage is *wide dilatation*, but this may occur either very rapidly, or only after a long period of *contraction*. In the able paper of Dr. Ogston,* upon alcohol poisoning, as observed in the practice of a police-station, it is reported that of twenty six cases observed during the later stages, twenty had the pupil dilated, in the remainder it was contracted; but the author does not state clearly whether the contraction persisted throughout in those cases in which death occurred, or a very great depth of narcosis was reached. My own belief is, that it is a most rare event (if, indeed, it ever occur) for the pupil to remain contracted up to the time of death in fatal cases, or during the stage of stertorous breathing even in patients who subsequently recover. I have myself witnessed four deaths from poisoning by alcohol, and in all these the condition was one of wide dilatation for a long time previous to the fatal termination. And besides this, I have seen a considerable number of cases which were only with great difficulty saved, and I cannot remember a single case in which actual stertorous breathing existed for any length of time, and yet contraction of the pupil persisted throughout. MM. Lallemand, Duroy, and Perrin observed a curious fact in regard to alcoholic poisoning in dogs, viz.—that if, when the animal had already reached the stage of dilated pupil, a fresh dose were administered, momentary contraction occurred, soon followed by renewed dilatation. According to received theories of the action of narcotics, the contraction would be accounted for by the stimulant effects which predominate in the early part of the action of alcohol, and the dilatation which follows to the subsequent "reaction." I explain the symptoms differently. The contraction momentarily induced was, I believe, owing to a sudden injection of the iris with blood, from the "*pumping*" action of the heart, which would quickly follow the administration of the additional dose. The subsequent dilatation represented the expulsion of the blood from the vessels of the iris, which would occur as soon as the more complete death of the pupil-nerves had removed all nervous influence, and the commencing death of the tissues of the vessels themselves allowed a process similar to the formation of *rigor mortis* to begin in them.

(e) There is yet another effect of alcohol which I am inclined to ascribe to its paralyzing action on the sympathetic nervous system, viz.—the increased secretion of urine which, in almost all persons, is caused by large doses of any alcoholic liquor. So far as my own observation extends, this effect is produced in nearly all subjects to a greater or less degree, but the extent to which it is carried varies, not only according to the individual's natural susceptibility to diuretic influences, but also, as might be expected, in inverse proportion to the activity of the respiratory process and of the perspiratory function of the skin. It is not necessary to dwell on this matter at present, as we must return to it when speaking of the elimination of alcohol.

A general review of the phenomena of alcohol-narcosis enables us to come to one distinct conclusion, the importance of which appears to be very great. Namely, that (as in the case of chloroform and æther) the symptoms which are so commonly described as evidences of excitement, depending on a *stimulation* of the nervous system preliminary to the occurrence of narcosis, are in reality an essential part of the narcotic—that is the paralytic—phenomena. A "pre-anæsthetic" stage may indeed be observed, except in cases of very rapid saturation of the blood with a large dose of alcohol, but it is not marked by any such symptoms as are now referred to; it is already at an end when they appear. It will be described hereafter, under the title of "True Alcoholic Stimulation."

As we have now discussed the phenomena of the induction of alcoholic-narcosis, it seems proper, at this point, to turn our attention to the subject of elimination—to the means by which the system rids itself of the poisonous dose. The con-

sideration of this matter will necessarily involve an examination, to some extent, of the question whether the same sort
of process is set up by non-narcotic doses, and so far is an
anticipation of what must come later, but it would be difficult
to avoid this amount of confusion.

2. *The Elimination of Alcohol.*

It has long been known that, to a certain extent, alcohol
escapes unchanged in the breath. It had been asserted, but
not generally believed, that a small proportion, also unchanged, always passed out of the body in the urine and bile.
It had been proved long ago, by Percy, that a portion at
least of the alcohol, taken into the body, was to be found unaltered in the brain many hours after the dose had been taken.
It was commonly believed, however, that the greater part of
the alcohol ingested disappeared within the organism, and
only entered the excretions under some altered form.

It is hardly necessary to recall the explanation which the
majority of physiologists agreed to give of these phenomena.
The reader is familiar with the celebrated classification of
Liebig, in which alcohol assumed a definite place among foods
of the combustible kind, and was ranked as a heat-forming
aliment, capable of replacing the oily, starchy, and saccharine
materials of alimentation. According to this theory the union
of alcohol with oxygen, within the body, gave rise to the
formation of carbonic acid and water, and the generation of
heat. The subsequent researches of Bouchardat and Sandras,
and of Duchek, though much criticised, appeared to elucidate
the stages of the chemical process by which the transformation
of alcohol into carbonic acid and water was accomplished, and
to show that, previously to the final change, there was a
formation of intermediate compounds (aldehyde, acetic acid,
oxalic acid). The impression is still fresh, which was created
some three or four years ago, by the publication of the able
researches of Dr. E. Smith, and those of MM. Lallemand,

Duroy, and Perrin, by which the whole of this fabric of theory appeared to be destroyed. The latter authors stated positively that alcohol was not, in any sense, a food; that it was neither transformed nor destroyed within the organism, but re-, appeared within a comparatively short time in the excretions being by them eliminated "en totalité et en nature." In a series of elaborate experiments they appeared to prove that appreciable quantities of alcohol *always* begin to pass off by the skin, the lungs, and the kidneys, within a very short time after the dose has been taken into the *stomach*, and continues so to pass for several hours. They failed, after repeated attempts, to discover the intermediate compounds into which alcohol had been represented as transforming itself before its final change; and, on the other hand, they detected *unchanged* alcohol everywhere in the body hours after it had been taken; they found the substance in the blood, and in all the tissues, but especially in the brain and the nervous centers generally, and in the liver.

From this imposing mass of evidence, MM. Lallemand, Duroy, and Perrin concluded that the entire expulsion of all the alcohol taken into the body, in an unchanged form and within a short time, is certain, and that except indirectly (by modifying digestion) that substance has no alimentary properties.

The great amount of labor and time requisite for the performance of any experiments capable of testing the value of these researches will, doubtless, account for the fact, that till quite lately no formal examination of MM.' Lallemand and Perrin's investigations had appeared. It was at once objected to them, however, that they were defective in two distinct respects: (*a*) inasmuch as the doses given were always *intoxicating* doses, and (*b*) because the amount of alcohol actually recovered in any of the experiments was really very small. There was a widespread feeling of distrust and doubt as to the conclusions which were considered by the French observers to result from the facility with which unchanged alcohol can be detected in the excretions by the chromic acid test, especially as the precise value and degree of delicacy of the latter had not been ascertained. At length, however, M. Edmond Baudot has put on record* a carefully conducted series of experiments by which great doubt is thrown on the correctness of the opinions in question, especially as regards the part supposed to be taken by the kidneys in the work of elimination.

M. Baudot justly observes, that if the opinion of MM. Lallemand, Duroy, and Perrin be correct, a *sensible* result should appear on examination of the urine voided during the twenty-four hours immediately succeeding the taking of a dose of alcohol ; that is to say, a result capable of being distinguished by the *alcoometer*. In a series of more than twenty experiments he abundantly proves that such is *not* the case, the results of this mode of testing being absolutely *nil*, except when the dose of alcohol has been very excessive; in the latter case a perceptible effect is produced. In a further series of experiments he applies to the distilled urines which have been vainly tested by the alcoometer, the chromic acid test, and obtains decided indications of the presence of alcohol, either immediately or after a longer or shorter time. He then examines the *delicacy* of the chromic acid test, and believes that he discovers it to be capable of revealing the presence, in the urine, of a proportion of alcohol so small as *one centigramme to the litre* (·155 grains of alcohol to nearly a quart of urine). Now, as the total amount of urine passed during the twenty-four hours following even a strongly diuretic dose of alcohol would be, probably, not more than two or two and a half litres, we are entitled to conclude from M. Baudot's results, if they are correct, that the justice of MM. Lallemand, Duroy, and Perrin's conclusions is greatly invalidated. I have, therefore, reinvestigated this question with some minuteness.

Experiment XIX.—A wide test-tube, fitted with a cork

* Union Médicale, Septembre et Novembre, 1863.

pierced with two apertures, was half filled with alcohol (of Sp. gr. ·795), and carefully weighed. Two glass tubes were then adapted to it, one of which was fitted to an india-rubber tube, five feet in length, which was connected with a powerful and continuously acting bellows; the other passed through a tight-fitting cork into a U-shaped tube, the curved portion of which contained a measured quantity of a solution of one part of bichromate of potash in 300 of pure sulphuric acid. The bellows being put in action, a current of air was conveyed through the alcohol into the test-fluid. In less than seven minutes the whole of this fluid (7·5 cubic centimetres) was changed to a bright emerald-green color, which became more decided after the tube had been removed and allowed to stand for an hour. On weighing the test-tube, containing the alcohol, it was found to have lost not quite $\frac{8}{10}$ths of a grain. (These weighing operations were kindly performed by Dr. A. Dupré, with a very delicate balance.) It was, therefore, calculated that $\frac{1}{10}$th of a grain of alcohol had been employed in coloring each cubic centimetre of the test-liquid.

It was obvious, however, that this experiment only roughly represented the delicacy of the chromic acid test. The research was, therefore, continued.

Experiment XX.—Ten grains of Sp. vin. rect. (P. L.) were added to five ounces two drachms of distilled water. Of this mixture one cubic centimetre was added to an equal quantity of the chromic acid solution above described. An instantaneous change to pale emerald-green took place. We thus perceive that $\frac{3}{25}$ths of a grain of alcohol colored *immediately* a cubic centimetre of the test-liquid.

Experiment XXI.—The same result was obtained with an alcoholic solution half the strength of that employed in the last experiment. The change was instantaneous.

Experiment XXII.—Ten grains of Sp. vin. rect. (P. L.) were added to five ounces two drachms of distilled water. Of this mixture five minims were placed in a minute test-tube connected by caoutchouc with a glass tube bent at right

24

angles and dipping into one cubic centimetre of the chromic acid solution above described, in **the** bottom **of** a small test-tube. The tube containing the alcoholic mixture was now placed in a water-bath, the temperature of which was gradually raised to 200° Fahr. In a little more than five minutes the first change was observed in the color of the test-liquid. At the end of ten minutes it was of a dark brown color; at the end of fifteen minutes it was slightly changing to green; and on the application of gentle heat it immediately assumed a decided, though pale, emerald-green tinge. Here we have $\frac{1}{35}$th of a grain of alcohol, at the utmost, producing a decided effect upon a cubic centimetre of the chromic acid solution; and we have evidence that an aqueous solution charged with no more than two grains (and a fraction) of alcohol to the ounce would operate most decidedly upon the test.

These effects, however, are trifling when compared with those which may be obtained by allowing the test-liquid to stand aside, in a warm place, corked, for some hours after the alcohol has passed into it. Even half an hour will sometimes make the difference between an absence of perceptible effect and a distinct coloration with the characteristic emerald-green. The sudden application of heat to the test-liquor, or (what amounts to the same thing) the addition of water, which by its combination with the acid generates heat, often brings out the color at once.

The comparatively simple mode of distilling the alcohol from a suspected fluid into the test-liquid which was used in the last described experiment is, doubtless, a wasteful one, more or less alcohol escaping the action of the test. That this is the case may be judged from the following experiment.

Experiment XXIII.—Five grains of Sp. vin. rect. (P. L.) were diffused through twenty ounces of distilled water. One cubic centimetre of this fluid was added to an equal quantity of the test-liquid; the result was at first apparently *nil*. The changes of color, however, commenced in a few minutes, and in about a quarter of an hour a pale emerald-green was developed. Here $\frac{1}{120}$th of a grain of alcohol affected one cubic centimetre of the test.

Having now assured myself of the great delicacy of the test, I proceeded to make observations on urine.

Experiment XXIV.—A man, æt. thirty, in good health, having taken (at 2 P.M.) one pint of light beer containing about three per cent. of alcohol, the following observations were made. The first urine was passed at 5.10; quantity, twelve ounces; sp. gr. 1018. Two ounces of this fluid were concentrated by two distillations to about one-fourth their bulk. This concentrated fluid was then placed in an apparatus similar to that used in Experiment XIX., a current of air being driven through it, by a constantly acting bellows, into the test-liquid. To the end of the U tube was attached a second tube, containing one cubic centimetere of the test-fluid, intended to absorb the alcohol which might escape that in the U tube. After the apparatus had been working for half an hour, the liquid (7·5 cent. cub.) in the first tube was of a pale emerald-green color; that in the second was un-affected. No further development of the color of the first, and no change whatever in that of the second, occurred, though the apparatus was kept working for an hour longer. The 7·5 cubic centimetres of test-liquid, which had been changed to pale green, were placed in a test-tube side by side with another tube containing an equal quantity of the *pure* test-liquid. To the latter was added, drop by drop, a mixture of alcohol and water, 1 part to 249. When thirty drops had been added the fluid instantaneously passed from the preliminary brown stage to the most developed dark emerald-green, greatly exceeding that produced by the urine.

The second quantity of urine was passed at 6.10; it was four ounces in quantity, and, when treated in the same way as the first portion, produced not the least effect on the test.

Experiment XXV.—One glass of sherry (about two ounces) having been taken, by the same person, at 1.30 P.M., the first urine was passed at 4.20 P.M.; sp. gr. 1020; quantity, seven

and a quarter ounces. Half of this was concentrated to a
fourth of its own bulk, and dealt with in the way described
in the account of the last experiment. The 7·5 cubic centimetres of test never became emerald-green, but only a pale
shade of that color, and even this took forty minutes to develop. The liquor in the second tube was unaffected.

The second urine, passed at 5.58 (quantity, six ounces; sp.
gr. 1022), was entirely without influence on the test.

It must be noted that these observations were made in
circumstances precisely the most calculated to exaggerate the
eliminatory action of the kidneys, viz. in very cold weather,
during sedentary work in a very cold room. The lungs and
skin *must* therefore have been restrained from doing anything considerable toward elimination, and nearly, if not
quite all, the alcohol eliminated must necessarily have passed
by the kidneys, the principal eliminatory organs in *all* cases
according to MM. Lallemand, Duroy, and Perrin. But it is
obvious that the amount passed by the kidneys was fractional
only, and represented no considerable part of the alcohol absorbed, which, as the wine or beer was taken with but a light
meal of food (lunch), probably included all that had been
taken into the stomach (about ℥ss in each case).

The elimination of alcohol in the breath next requires consideration. We can easily assure ourselves that portions of
alcohol do escape from the lungs, even after a moderate dose
only has been taken. Preliminary experiments convinced me
that this is always the case: the fact of elimination may be
ascertained, within eight or ten minutes after taking a single
glass of Bass' ale, for instance, by directing the current of
expired air through a portion of the test-solution of chromic
acid. The test experiments were made in the following manner: A mouthpiece of vulcanized india-rubber was fitted
carefully over the mouth (leaving the nostrils free); this communicated with a glass tube which pierced the cork of a U-
shaped tube open at its further end. The curve of the U
tube was occupied by 7·5 cubic centimetres of the test, through

which the breath was made to pass. Inspiration was performed through the nose, which was carefully held closed in expiration. Operating in this way, I found that the test-liquid became changed to a decided emerald-green in from ten minutes to a quarter of an hour after he had taken a moderate dose of alcohol such as has been mentioned; the change *commenced* in about five minutes.

In order to form some idea of the extent to which elimination by the lungs proceeds, the following experiment was made:

Experiment XXVI.—Half a pint of Bass' bitter ale having been taken at 2.10 P.M., expiration through test-liquid (7·5 cubic centimetres) was commenced within four minutes afterward. The U tube was made to communicate with a test-tube containing one cubic centimetre of the chromic acid, which was intended to catch any alcohol which might escape the first portion of the test. Expiration through the apparatus was steadily carried on till 3.35; at the end of this time the liquid in the U tube was of a dark transparent emerald color, the liquid in the second tube a very pale shade of the same tint. On removing the colored liquid and charging the U tube with 7·5 cubic centimetres of fresh test-liquid, no change was produced, although the operation was continued for an hour. The original charge of the U tube was now placed in one test-tube, and an equal quantity of fresh test-liquid in another, and to this last five minims of absolute alcohol were added, *guttatim*. Instantaneously a change was produced in the liquid, which in two minutes and a half had turned the liquid to a dark emerald-green, undistinguishable from that colored by the breath. It would therefore appear that the effects noted in the experiment *might* have been caused by an elimination from the lungs of a total quantity of not more than five minims of alcohol. In order to ascertain whether a considerable quantity of alcohol had escaped by the kidneys, the whole of the urine passed during the twelve hours next succeeding the dose was collected and examined. It consisted of two portions: 6 ozs. had been passed three hours and a half after the dose of alcohol, and 8¾ ozs. passed about seven hours later; half of the first portion, concentrated by distillation to 1 oz., was placed in a flask and immersed in a water-bath, of which the temperature was gradually raised to 200° Fahr. The vapor was carried by means of tubes into two successive portions (one cubic centimetre each) of the test-liquid. This process was carried on gently for an hour: at the end of this time, the liquid in the first tube was colored dark emerald-green; that in the second tube was a very light shade of the same color. The change of color had not commenced, even in the first tube, till the vaporization had been going on for more than ten minutes. These effects are certainly not those which would have been observed had a large proportion of alcohol been eliminated by the breath and urine, but it appeared that they represented the whole of its action (barring those portions wasted in the experiment, which I believe were inconsiderable), for on breathing through a fresh portion of test-liquid no effect was obtained; also, the second portion of urine yielded no result, either by direct addition of the chromic acid to it, or by the use of this test after distillation. Without pretending to any minute quantitative exactness of statement, we are certainly entitled, I think, to regard the elimination which took place by lung and kidney, in the above experiment, as trifling in extent. Unless we could believe that a large portion had escaped by the skin, it is obvious that but a fraction of the alcohol which entered the body was eliminated from it during the hours immediately succeeding the dose.

The following experiments were made in order to ascertain the length of time during which elimination by the lungs continues, when a large dose has been taken. The apparatus containing the test-solution was a U-shaped tube, the curve of which was occupied by 7·5 cubic centimetres of the chromic acid solution: a second tube, containing one cubic centimetre of the test, was connected with this, in order that as little alcohol as possible might escape. The experiments were very unpleasant to the patient, causing headache and a dazed confused feeling, but no absolute loss of intelligence. The rpidity with which these large doses were taken would rather add to the probability of **a copious and rapid** elimination.

Time of administration, 10 P.M. to 10.30.	℥iij Sp. vin. rect.	℥iij whisky.	℥iij brandy.	
Examination commenced at 12 P.M.	Solution in U tube colored dark emerald in 7 minutes.	Solution in U tube colored dark emerald in 5 minutes.	Solution in U tube colored dark emerald in 8½ minutes.	Solution in 2d tube, pale emerald in all three cases.
Examination commenced at 2.30 A.M.	Solution in U tube colored dark emerald in 15 minutes.	Solution in U tube colored dark emerald in 12 minutes.	Solution in U tube colored dark emerald in 21 minutes.	Solution in 2d tube, pale emerald in Exps. 28 and 30; unchanged in Exp. 29.
Examination commenced at 9 A.M.	No effect.	No effect.	No effect.	

In all three cases the morning urine was tested, **by adding** to it **an equal bulk** of chromic acid solution, and **a decided** effect was produced. Four **hours later, a freshly secreted portion** of urine still produced a faint reaction with the chromic acid test, in Cases 28 and **29.**

I have not had the opportunity, so far, of examining the phenomena of elimination **of alcohol by the skin,** except in a tentative manner, which has sufficed **to convince** me **that** such elimination always, or nearly always, takes place; **or,** at least, that this peculiar change of color may nearly always be produced in the chromic acid solution by passing **the exhalations** from the skin through it, after taking alcohol **in any dose.** But I cannot think this is of much consequence **to** the general question. It is not pretended, so far as **I am** aware, that elimination by the skin would dispose of more than a very small fraction of the alcohol taken into the body, except possibly, in the case of unusually violent perspiration, from external heat or muscular exercise. It would certainly be very **unreasonable**, in the event of elimination by the lungs and kidney being shown to be obviously unequal to the task of disposing of more than au inconsiderable fraction of the alcohol taken into the body, to attempt to account for the remainder by the theory that it passes through the skin.

That the elimination effected by the lungs and kidneys *is* entirely disproportionate to the quantity of alcohol taken into the system we can hardly doubt, when we review the experiments recorded above, in conjunction with those of M. Baudot. Singularly enough, some very strong arguments in favor of this view are furnished by the authors of the doctrine of entire elimination. Nothing is more plainly proved by MM. Lallemand, Duroy, and Perrin, than the fact that long after the latest periods at which any of the alcohol absorbed can be recognized in the breath, the urine, or the sweat, unchanged alcohol in notable quantities can be recognized in the blood and tissues of the alcoholized animal. M. Baudot justly observes that there is no necessity to suppose that this substance must be transformed *immediately*, if transformed at all, in the organism. It may well be, for all we know, that long after the elimination of that small surplusage which we are able to detect in the urine, breath, and perspiration, during a few hours, the remainder very slowly undergoes a change which as yet has **not** been satisfactorily traced ; and it is certain that the experiments of MM. Lallemand, Duroy, and Perrin do not in the least provide **against this** objection, since in none of them were the blood and tissues examined **at** a sufficiently late period. A **very** important fact educed by these authors tells strongly in favor **of the idea of** a series of transformations. They do, indeed, **fail** to find proofs of the transformation of alcohol into aldehyde, **but** on administering the latter substance to animals they find *acetic acid* in the blood (only two hours after the administration) along with a certain quantity of untransformed aldehyde. This fact can hardly fail to suggest the probability that more careful and extended research may one day discover that alcohol is transformed into aldehyde, although the latter change may not occur so rapidly as the oxidation of aldehyde into acetic acid.

That a sensible proportion of alcohol, such as may even be recovered by distillation and identified by its character of inflammability, should escape in the urine when excessive doses have been taken, is by no means surprising. Owing, probably, to its paralyzing influence upon the sympathetic nerves, this substance, in narcotic doses, rarely fails to produce a considerable diuresis: this fact has been recognized by all observers. As alcohol is freely soluble in water, it is natural that a sensible proportion should pass out with the kidney-secretion; and it has been remarked by more than one author that it is probably in this way that certain persons escape intoxication (in the ordinary sense of the word) after large doses of alcoholic liquors. And that minute quantities should be eliminated by the kidney for a few hours immediately succeeding even a moderate ingestion is only what we might expect, now that we have reason to believe that the blood contains unchanged alcohol for so considerable a period after it first receives that substance; under such circumstances, the urine could hardly fail to contain a small portion.

It must be understood that what has now been stated as to elimination is so stated conditionally only. I have assumed, in deference to what seems the general opinion, that a volatile product which has been obtained from the urine, breath, &c., and which causes the characteristic changes in the chromic acid solution, must, of necessity, be either alcohol or one of a few allied substances, the absence of which may be predicated with certainty in ordinary cases. I have myself no fair grounds for challenging this view, and have therefore not attempted to make capital by doing so; at the same time, it is right to mention that several able chemists have expressed to me a considerable doubt whether some of the results of alcoholic transformation may not equally affect the test, notwithstanding the assertions of MM. Lallemand, Duroy, and Perrin, supported to a certain extent by proofs, to the contrary effect.

It is right, also, that I should point out the difficulty and uncertainty which attends any attempt to judge *color by the eye* with anything like quantitative exactness. This difficulty is too much ignored, I think, by MM. Lallemand, Duroy, and Perrin.

PART II.

True Alcoholic Stimulation ; or, the Non-Narcotic Action of Alcohol.

THE subject which we now approach is one which would require a large volume to do it full justice. I must content myself with a few striking illustrations, which will exhibit, from different points of view, the reality and importance of this phase of the action of alcohol, the very existence of which has been and is denied by many.

a. The first examples which I shall produce are found in the effects of small, as contrasted with those of large, doses upon various convulsive disorders.

I have already spoken* of the powerful effect which alcohol often exerts in averting threatened *epileptiform attacks*, and have insisted upon the fact that it is a small dose only which is required, an excessive quantity being neither necessary nor safe. I may now state that, altogether, I have seen seven epileptics who with greater or less regularity had employed this remedy and were aware of its power : for my own part, I confess that I hesitate to order its employment by any but a person of great intelligence and firmness of mind, on account of the almost invincible tendency which patients have to overdo any remedy which they fancy will avert some terrible danger —a measure which, in the case of alcohol, might have the disastrous effect of leading to drinking habits. It would be a great mistake to suppose that drowning a patient in drink will dissipate the convulsive tendency; on the contrary, it is well known that some epileptics who are tolerably free from

* *Vide* Chapter III.

their convulsive attacks except when they take so much liquor
as to become nearly comatose, on the induction of the latter
state are certain to experience a seizure.

1. The effect of alcohol in arresting the *convulsions of teething* is one of the most remarkable instances of a real therapeutic influence which can be witnessed. For my own part,
there is no other plan of treatment from which I have seen
such benefit produced as has resulted from this; and I may
fairly say that I have seen every kind of treatment. There
is not the least necessity for intoxicating the little patients;
a minute dose of wine or brandy (for young infants a few
drops at a time in a little water) is amply sufficient for any
good purpose that can be effected. Under my own hands
this plan has been most successful in the few cases in which
I have as yet been able to adopt it, but I have heard from
other practitioners of a really extensive trial of it with most
excellent results.

Excellent results have been obtained, also, in some cases of
tetanus, by the use of alcohol; the presence of this disease
seems to have an influence (like that of fever, to be referred
to presently) in preventing the ordinary poisonous effects of
very considerable doses.

The importance of this distinction between the influence of
large and small doses, respectively, on convulsive affections,
if it were real, has led me to make the following researches
on the action of alcohol in the convulsions of narcotic poisoning.

Experiment XXVII.—Mr. Squire's preparation of the
bimeconate of morphia is well known to be an excellent
anodyne and soporific, producing, in moderate doses, a com-
forting and soothing effect without exerting any true narcotic
action. It occurred to me to try whether the usual tetanic
convulsions induced in frogs by opium would follow the administration of a poisonous dose of this preparation. Accordingly, I injected into the cavity of a frog's body twenty minims
of Squire's solution. In six minutes the animal was completely tetanized, and in fourteen minutes it was dead. I

thought this was remarkable, because, though tetanus is not an uncommon result of poisoning with preparations of morphia in animals, it is seldom so violent or so rapidly fatal as this.

Experiment XXVIII.—Seventeen minims of solution of the bimeconate, deprived of nearly all its alcohol by evaporation under a current of air aided by gentle heat, were injected into the belly of a frog, as nearly as possible equal in size and activity to the subject of Experiment XXVII. The animal rapidly became narcotized, and, at the end of five minutes, was completely palsied and insensible. In ten minutes he was dead, no tetanic symptoms having occurred, but rigor mortis very soon set in.

Experiment XXIX.—One grain of opium was dissolved in one drachm of distilled water: of this solution twenty minims were injected into the belly of a frog. Tetanic spasms commenced in seven minutes, and eight minutes later the animal died. Rigor mortis set in immediately.

Experiment XXX.—Twenty minims of the same aqueous solution of opium, with one minim of Sp. vin. rect., were injected into the belly of a frog. In five minutes the animal was paralyzed and insensible, and in this condition he remained for forty-eight minutes, and then died. No tetanic spasms had occurred, and the muscles remained lax for some time after death.

Experiment XXXI.—Twenty minims of the same aqueous solution of opium, and four minims of Sp. vin. rect., were injected into the belly of a frog. Tetanic spasms supervened in eight minutes, and the animal died ten minutes later.

Experiment XXXII.—Twenty minims of the same aqueous solution of opium were injected into the belly of a frog, together with fifteen minims of Sp. vin. rect. No tetanic spasms took place, but the animal became immediately paralyzed and unconscious, and respiration ceased in three minutes. Rigor mortis instantly followed.

Similar observations were made on the modifying influence of alcohol in strychnia poisoning in frogs. The results were the same. Alcohol is obviously not directly antidotal to tetanizing agents, but as a narcotic it may paralyze, or as a stimulant it may temporarily arrest their toxic action altogether.

2. The anodyne and soporific influence of doses of alcohol insufficient to narcotize affords us another illustration of the true stimulant action of this substance. I have already alluded to this subject (see Chapter III), and have remarked that it is not by intoxicating doses that these objects are best attained. It is true that it is possible to quiet pain, and to throw the patient into a state of stupor which somewhat *resembles* natural sleep, by the use of intoxicating doses of alcohol, but this is not the kind of effect that should be sought for. The object of the physician is simply to restore the natural condition of the nervous system, not to add to its disturbance by inducing narcotic poisoning.

In no circumstances is the action we are now considering better illustrated than in certain forms of rheumatic pericarditis, viz. those which are distinguished by great pain, sleeplessness, and jactitating movements of the limbs. There are many cases in which the use of opium is contraindicated by special circumstances, and there are others in which that drug proves ineffective. When this is the case, and especially when with this there is an inability to take beef-tea and the like, nothing acts so favorably as alcohol in repeated small doses, the production of even the minor signs of intoxication being carefully avoided. Of the two, alcohol is, in my experience, superior to opium as an anodyne and soporific in this form of disease.

In cases of acute neuralgia it has happened to me several times to observe, that after large doses of various narcotics had been tried in vain, the first real and substantial relief was obtained by the use of a moderate dose of alcohol; and I have more than once experienced this kind of effect in my own person when tormented with an unusually severe attack of neuralgia of the fifth nerve. The consequence of overdoing this remedy is, however, nearly always disastrous. The patient

awakes from the heavy stupor (rather than sleep) into which
he has been thrown, relieved indeed from the neuralgia, but
tormented with the dull diffused headache of receding narcosis,* pale, trembling, nauseated, and in a condition highly
favorable to the recurrence of the neuralgia in a severe form.
It is obvious, therefore, that if alcohol is to be administered
at all for the relief of neuralgia, it should be given with as
much precision, as to dose, as we should use in giving an
acknowledged deadly poison.

The classical illustration, however, of the favorable soporific
influence of alcohol, is to be found in its use in low fevers, such
as typhus and typhoid. Given a certain rapidity of pulse, we
may nearly always assure ourselves, in cases of these diseases,
that the patient will be unable to obtain natural sleep, but in
place of this will pass into a state either of coma or delirium.
Under such circumstances, it too often happens that meat,
broth, &c., cannot be retained by the stomach; and it may
also happen that they be retained by the rectum. Notwithstanding all that has been said to the contrary by persons
who have never fairly tried the plan, there is nothing which
meets the exigencies of this condition with an efficiency which
at all approaches that of alcohol administered in repeated
non-narcotic doses. The strongest proof that this is the case,
and that the action of the remedy is salutary, is to be found
in the fact that large doses may be given without producing
any narcotic effect. It has been in vain attempted to explain
this fact away. It has been ingeniously suggested by Dr.
Edward Smith,† for example, that a large portion of the
alcohol, in such cases, altogether fails to be absorbed, and is
ultimately evacuated with the dejections. I believe that few
persons who have served the office of clinical clerk in a hospital where fever cases are frequently received, and alcohol is
freely used, will consent to allow this view. I have inspected

---

* See case at page 121.

† On the Mode of Action of Alcohol in the Treatment of Disease. Trans. Med. Soc.
of London. 1861.

the dejections of a large number of fever patients, and, except in a few cases attended with very severe diarrhœa, have never recognized the odor of alcohol in them. On the other hand, I have directly experimented on the effect of adding as small a quantity as half a drachm of brandy to an ordinary typhoid dejection, and have found the odor perfectly distinguishable. May it be that the alcoholic liquors become changed to acetic acid in the stomach, and are in that form passed with the dejections? I believe this is also impossible; because the presence of acetic acid, or of acetates in any considerable quantity, in the fæces, would attract attention from their vinegary odor.

The patient to whose dejections brandy was experimentally added, as above described (a girl of thirteen years), had been taking twelve ounces of brandy per diem for three days previous to the experiment, without the slightest alcoholic odor being distinguishable in her stools.

Can it be that alcohol in unusually large proportion is eliminated by the lungs, skin, and breath, or by either of these channels? It is very easy to suggest that this may be the case, and it is very difficult absolutely to prove the negative; but the following arguments may be adduced against such an idea. With regard to the lungs, it is notorious to experienced hospital-nurses that an excessive alcoholic odor in the breath is usually connected with the obvious disagreement of the remedy, and is accompanied by flushing of the face, and signs of genuine intoxication. It is scarcely proper to put a fever patient to the annoyance of breathing through an apparatus by which the alcohol might be collected, or passed into the chromic acid solution; but this proof is really not necessary to convince ourselves that the process of elimination by the lungs is not going on with anything like the rapidity with which it proceeds in persons who are recovering from drunkenness.

As to the kidneys I can speak more positively. The following experiments were made on the urine of two patients who were taking large quantities of alcohol with obvious benefit, and without the least symptom of intoxication.

Experiment XXXII.—A woman, æt. 36, suffering from typhus fever, with a well-developed rubeoloid rash, was seized, on the tenth day of the disease, with rather violent delirium; the skin was hot, and rather dry; pulse 146. Six ounces of brandy per diem had been allowed so far: the quantity was now doubled. During the ensuing twenty-four hours the pulse came down to 120, the delirium ceased, and the tongue became less dry than it had previously been. The whole of the urine passed during this time was collected: quantity, 42 ounces; specific gravity, 1015; reaction acid; on the application of heat and nitric acid, a minute quantity of albumen was thrown down. No alcoholic odor could be perceived. Four ounces of this urine were concentrated by distillation till one ounce and a half of clear liquid, possessing a slight animal odor, with a suspicion of the smell of alcohol, was obtained. One drachm of this fluid was redistilled* on a water-bath, a carbonate of potash tube being interposed between retort and receiver in order that as much aqueous vapor as possible might be arrested. Ten minims of clear fluid were obtained, with a distinct alcoholic smell, but which refused to ignite. Here we may say roughly that the liquid finally obtained represented a condensation of the urine to $\frac{1}{18}$th of its bulk: we get the result, then, that, if the whole urine had been concentrated to ℥ij ʒv (or therabouts), it would still have been uninflammable, i.e. must have contained less than 50 per cent. of alcohol—less than ʒxss, that is. In reality, however, it probably contained very much less; for, on diluting the ten minims of liquid obtained by the last distillation with 19 parts of distilled water, and adding a cubic centimetre of this fluid to an equal bulk of the chromic acid test, scarcely any change in the color of the latter was pro-

---

* In this miniature distillation, the retort and receiver were respectively represented by small test-tubes. The receiver was plunged in ice-cold water, and was loosely corked, the cork being pierced by the drawn-out nozzle of the drying-tube.

25

duced, even after it had stood for an hour. (Compare Experiment XXIII.)

Experiment XXXIII.—A man, aged 43, suffering from severe erysipelas of the head and face, was taking twenty-four ounces of brandy per diem.   On the second day of this treatment the whole urine (51 ounces) was collected.   Specific gravity, 1021; acid reaction; no albumen.   Two ounces of the urine were condensed by distillation until a little more than half an ounce of a clear liquid was obtained, smelling slightly of alcohol, and strongly of some animal matter.   One drachm of this fluid was redistilled over a water-bath, the vapor being carried through a drying tube, and about twelve minims were received in a test-tube.   This liquid refused to ignite; it was diluted with nineteen parts of distilled water; one cubic centimetre of the dilute fluid produced only a very slight change in an equal quantity of the chromic acid test, after many minutes.   Here the successive concentrations may be said to have reduced the urine operated on to $\frac{1}{25}$th of its bulk; yet the product was uninflammable: that is to say, the whole urine must have contained less than 3x of alcohol, and probably contained *very much less*.

In both these experiments (XXVII. and XXVIII.), the original distillates were also tested by adding dried carbonate of potash: only a very small layer of alcohol separated, sufficient, when drawn off by a pipette, to moisten a lamp-wick, which burned freely.

Both these experiments are doubtless open to the objection that there may have been waste of alcohol in the successive distillations.   From the precautions adopted, however, I do not believe this amounted to more than a very small fraction. In every such operation water nearly ice cold surrounded the receiver.   When all proper deductions have been made on this score, the fact still remains that the quantity of alcohol eliminated by the urine is entirely insufficient to explain the absence of symptoms of ebriety in the patient.

With regard to the possibility of the skin assuming an un-

usually vigorous eliminatory action in these and similar cases,
I certainly do not believe that this occurred. I had not the
means of testing the matter otherwise than approximatively;
but this is the ground of my opinion. In neither case was
an alcoholic smell of the skin to be detected. Now, I have
been at the pains from time to time of examining the skin of
a very large number of drunken people, and I have never
once failed to notice a perceptible alcoholic odor proceeding
from them. (It is very easy to persuade one's self erroneously
that there is a smell of alcohol from the skin of a fever
patient, because there is nearly always some remnant of
brandy or wine standing in a glass not far from his bedside:
this source of mistake must be carefully avoided.) It is not
even necessary that a great excess should have been committed; a very slight amount of intoxication is sufficient, with
most persons, to insure the occurrence of this symptom.

The same answer may be given to the suggestion, which
will probably be made, that the alcohol may have been chiefly
eliminated in the breath. I do not believe this is possible,
because the breath of a person who has really committed an
alcoholic excess, such as to cause the slightest symptoms of
intoxication, is almost certain to exhale a powerful alcoholic
odor: this may be concealed by the smell of other food which
has been taken, but in the absence of such a source of confusion it may always be detected. Whereas the breath of
these patients, and that of numerous others to whom alcohol
has been given in large doses with great benefit in similar
circumstances, hardly betrayed more trace of alcohol than an
experienced nose will recognize in the breath of a man who
has taken only a pint of beer. It is quite true that these are
inexact statements, and are not worthy to be dignified with
the name of *proofs*. I have no wish that they should be so
called, but merely that they should receive as much attention
as the similarly inexact observations of MM. Lallemand,
Duroy, and Perrin. Taken along with other facts which I
advance, they appear to me to incline the balance strongly in

favor of the belief that alcohol is retained within the system, more especially in fevers and other adynamic conditions of the nervous system, for a special purpose quite apart from that of merely aiding digestion. It certainly appears as if the influence it has in removing delirium, calming pain, and inducing natural sleep, were exerted in virtue of a power by which it assists, directly or indirectly, in the repair of the nervous system.

3. By far the most important aspect, however, of the non-narcotic action of alcohol is that which is presented to us by the singular cases which, however they may have been discredited, certainly do occasionally present themselves, of individuals subsisting for considerable periods of time without, or nearly without, other subsistence than alcohol, and yet escaping the prostration which entire starvation for a similar period would undoubtedly produce. The facts of this kind may be divided into two groups: (1.) Those which concern the support of the organism in acute disease. (2.) Those which concern the maintenance of bodily vigor upon an extremely insufficient allowance of ordinary food in a state of ordinary health.

(1.) As to the support of the organism during the progress of acute disease, it is not disputed by those who have given a proper trial to the treatment that, in the partial or total absence of the power to take other food, patients in typhus, pneumonia, &c., frequently maintain vital power and preserve their intellectual faculties throughout, and on the termination of the severe symptoms of the disorder, convalesce with remarkable rapidity. It is said, however, by Professor Beale, who certainly has had large experience of the alcohol treatment, that it is not in virtue of any really *alimentary* properties of alcohol that this is effected; the action of this substance, in his opinion, merely arrests the too rapid cell-changes going on; and as a proof that the organism has only subsisted upon the consumption of its own tissues, he states that the convalescence even of patients who have taken considerable quantities of beef-tea, &c., in addition to large quantities of alcohol, is attended with great emaciation. For my own part, I have always made the rule, and have seen no reason to depart from it, that large doses of alcohol are only to be employed when other nutriment cannot be taken or cannot be digested, except in small quantities, or when the emergency is sudden and we cannot wait for the action of ordinary foods. Acting on these principles, I have not found the convalescence of patients in acute disease attended with great wasting and debility; the very contrary, in fact, has been the case, and in no instances have I ever seen better and more rapid recovery of full health and strength (after an equally severe illness) than in the following cases, which occurred in my practice at the Chelsea Dispensary, and in which there was total inability to take *any* common food for days together.

Case I. Pericarditis occurring in the course of acute Rheumatism.—In February, 1861, I was called to see a young man, a plasterer, æt. eighteen, who was suffering from acute rheumatism. He had been treated for four days homœopathically. He suffered great pain from three or four inflamed joints and also in the chest; the pulse was 120, weak and irregular; the respiration 45; body bathed in sweat, jactitating movement of the arms, and slight delirium. On auscultation, I discovered a loud pericardial friction sound. Not a bit nor a drop of any food could be retained in the stomach, except cold water, in teaspoonfuls. The medicinal treatment in this case consisted solely in the administration of one grain of morphia every four hours. (This was continued throughout the persistence of the chest symptoms.) For seven days the man's only nutriment was 12 ounces of gin per diem and about an equal quantity of water (two drachms of the gin being given every half hour), and for some days longer but very little common food was taken, for the sole reason that gin and water was the only thing the stomach did not absolutely reject at once. Recovery in this case was very rapid, for the man was able to get about within

the month. It was interesting in this case to observe (a)
that there was never any inebriation, though the patient was
a sober lad, quite unused to drink spirits, (b) that the pulse
and breathing steadily came down under the treatment, (c)
that the recovery of the appetite for ordinary food, and the
recovery of muscular strength, was in a manner *sudden*, and
happened directly that the sickness ceased. The emaciation
was so trifling as not to deserve that name.

Case II.   Pneumonia.—I was sent for in February, 1860,
to see a child, æt. fourteen months, who was suffering from
severe pneumonia; pulse and breathing were very rapid, *skin
hot and dry*, cough distressing. The stomach would retain
no food, not even water when given alone. Port wine and
water the child drank greedily from a spoon : and on this
diet alone the child subsisted for twelve days, taking nearly
six ounces of wine a day. At the end of that time the stomach
was as irritable as ever to common food, but a little cod-liver
oil was taken, and agreed. Gradually the cod-liver oil was
altogether substituted for the wine; this occupied about ten
days more, and by this time a good natural appetite was beginning to show itself. After all this abstinence from common
food, the child, notwithstanding that it had been extremely
restless during the whole illness, recovered most rapidly, and
its emaciation was so trifling that it looked quite fairly plump,
considering it had never been very fat.

Case III.   Double pneumonia in an adult.—In July, 1859,
I attended a man, æt. twenty-four, who suffered from inflammation of both lungs, and whose case was a very bad one.
He was a strong young fellow naturally, but an attempt to
administer tartar emetic, a remedy which in such a case as
his one might naturally think proper, produced such severe
depression and nausea that it was at once given up. No
beef-tea, arrow-root, nor any other food except brandy, could
this patient take; twelve ounces of the latter per diem, and
ultimately double this quantity, were given, in divided doses,
every half hour; and for ten days he lived on *nothing* but

this and a little water. In a month from the time of my first
visit the man was well enough to resume his work. There
could be no mistake as to the fact that emaciation in this case
was trifling, this was not merely a conjecture from superficial
inspection, but was proved by the patient, who was lean and
muscular, and had scarcely any fatty tissues to be consumed,
recovering muscular power with such great rapidity.

Case IV. Pleurisy.—In March, 1860, I was called to see
a man, æt. seventy, who was suffering from acute pain in the
side and dyspnœa; the pain had lasted twenty-four hours;
respiration 45, pulse 130; skin hot and dry; loud friction
sound over the lower half of the antero-lateral portion of the
left chest. One grain of morphia every four hours (at first)
constituted the sole medicinal treatment in this case; for six
days no nutriment except ʒss of brandy and an equal quantity
of water every hour was administered, for the stomach would
retain nothing else. At the end of this period, the chief
symptoms being now on the mend, the appetite began to
return, and ordinary food was gradually substituted for the
alcohol. The same cessation of delirium, lowering of the pulse,
&c., which has been mentioned in the reports of some other
cases, was noted here.

Convalescence in this case was extremely rapid, and the
man had no sooner regained completely his appetite than he
was able to walk about in a manner truly surprising, considering his years and the severity of the attack. His own impression, and that of his friends, was that the amount of flesh
he had lost was almost *nil*.

The element of imperfection in the above cases is the
absence of any direct proof as to the trifling character of the
emaciation of the tissues, such as would have been afforded
by *weighing* the patients before and after this remarkable
abstinence from common food. Unfortunately, patients about
to be afflicted with acute disease seldom give us notice beforehand, so that we may put them into the scales and ascertain
their weight while in health! To reject the evidence afforded

by the remarkably rapid resumption of apparent plumpness, of bodily strength, and of intellectual and sensorial activity, which was noted in these cases, would, as it appears to me, be a dishonest straining at a gnat.

I could add to these cases the record of many others in which the amount of common nutriment taken was utterly insufficient to account for the conservation of the tissues which was observed, but it appears to me needless to do this, because the instances already quoted are crucial. It is a serious error to suppose that patients who struggle through an acute disease **by the unaided** powers of nature in the absence, partial or total, of nutriment, are able to make rapid **recoveries like those recorded above**: on this point I appeal to all those who have watched the convalescence of patients who have been lucky enough to escape with life after an acute pneumonia treated with *ptisanes* (of nothing particular) and *diète absolue*. Nature will put up with a good deal, but the combination of acute disease and starvation is really rather too much even for her patience.

But even were it necessary to surrender all claim on the part of alcohol to be reckoned as a conservator of the bodily frame generally, during acute disease, with regard to the nervous system, at least, I might safely challenge all comers to explain the phenomena observed in the treatment of acute diseases, otherwise than by attributing a food action to this substance. It·certainly has been suggested by Dr. Beale that alcohol chemically arrests the vital cell-growth in nervous tissue by coagulating its albumen (or by producing an effect approximating to this); and in this way prevents the too rapid waste of vital power. But there is no probability, I think, that alcohol, in the thousandfold dilution in which it reaches the brain, medulla oblongata or spinal cord, has any coagulating power, especially on tissue placed in the complex conditions of "life:" the speculation appears gratuitous, and *à priori* very improbable. Granting even that the excessive waste of the nervous system is the source of danger in typhus fever, how can we suppose that a mere coagulation or thickening of the albuminous matter of the brain and spinal cord is likely to arrest these changes? According **to Dr. Beale** it **would accomplish** this by inclosing the "germinal" matter in an impermeable coating, through which no nutriment could penetrate to it from without. But how are we to explain the effects of alcohol on this basis? How is it that it calms delirium and promotes natural sleep in a manner undistinguishable from that in which ordinary food, such as meat soup, acts? The effect of nutritious food, where it can be digested, is undistinguishable from that of alcohol upon the abnormal conditions of the nervous system which prevail in febrile diseases.

(2.) Passing from this subject, we have to consider **the extraordinary** way in which the healthy system (using this word comparatively) adapts itself, in some cases, to a diet composed chiefly, or almost entirely, of alcohol. Before entering on this matter, I desire to make one or two preliminary re**marks.** There is, of course, a great difficulty in understanding how such occurrences *can* take place, and equally **of** course there are a certain number of persons ready at once, and in the most candid manner, **to declare** their belief that they *never did* take place, and that any one reporting their occurrence must be either a dupe or an impostor. Such was **the burden, for** instance, of certain comments which were hastily made on some of the facts to be now related, when originally published in another shape. It had therefore been my desire to publish the histories of some cases with which I have been acquainted, which, if I were at liberty to repeat the professional names concerned, and to describe the way in which my information was obtained, would convince the most skeptical that there is no exaggeration in the statements now to be made. Unfortunately, but very naturally, there is a great reluctance, on the part of those immediately concerned, to allow even the most moderate publicity to be given to such facts; neither the toper himself nor his medical attendant

desire him to be made an example: otherwise the mere statement of certain names would be sufficient guarantee of the
accuracy of the present narrator. I address myself, therefore, to those who are willing to receive a carefully recorded
narrative, although it may not agree with their theories,
without at once supposing that it *must* be erroneous.

The case which first attracted my attention to the possibility of the human organism subsisting, under certain circumstances, for a long time upon alcohol only (or practically
so), was that of an old man who became my patient in 1861
at the Westminster Hospital; he was at that time eighty-three
years old, and was half led, half carried to the hospital by
his daughter for relief from chronic bronchitis, which was
just then severely aggravated. It happened that in prescribing for him, the subject of stimulants was mentioned, and
the old man, apprehending that as a matter of course I should
put him on a severe *régime* (he was an old soldier), begged
and prayed that his "drop of drink" might not be taken
away from him. I somewhat carelessly inquired what the
amount of the said "drop" might ordinarily be, and received
the astonishing answer—"A bottle of gin a day." Here the
daughter interfered, and remarked that it would be death
to him to take away this source of support, for he ate no food.
Inquiring what "no food" meant, I was assured, again and
again, that one small finger-length of bread, usually toasted,
was all that he *ever* took from one end of the day to the other;
he was occasionally drunk, but not often, and was a man of
astonishingly active habits for his age. I need hardly say
that it was a matter of utter impossibility for me to watch
this man and his friends day and night to make sure that
they were not deceiving me; but their story was confirmed
by a neighbor who brought the old man on one occasion. I
kept this case in view for a twelvemonth, when a renewed
access of bronchitis carried the patient off. Again I made
searching inquiries, and was assured that the man's habits of
life were such as I have mentioned, and that they had been

so for a great number of years, about twenty years his daughter believed. It was certain he did not take tea or coffee, or anything of that kind even; and the only thing besides his gin and water (which he drank unsweetened) which could have helped to support him was the daily fragment of bread and a few pipes of tobacco.* His very expensive manner of living prevented his friends, who otherwise seemed decent folk, from affording paid medical advice. The man's appearance was very singular and not easy to describe; it was not that he was very greatly emaciated, but he had a dried-up look which reminded me of that of opium-eaters.

Another case, the facts of which came to my knowledge, was that of a gentleman, a manufacturer, of active habits, and ordinarily leading a sober life, who was liable nevertheless to periodical outbreaks of something like insanity, in which he would drink for days or even weeks together. On these occasions it was repeatedly observed that he abstained altogether from solid food; he was known to do this for periods varying from a few days to as long as three weeks; and yet it was remarked by his friends that at the end of these bouts he never appeared emaciated, and that he resumed his ordinary occupations with scarcely any of the debility from which he might have been expected to suffer under the circumstances.

The following cases, which are similar to the above, are narrated by Dr. Inman, along with several others which are less decisive. Dr. Slack, of Liverpool, informed Dr. Inman, that two female patients of his own, who loathed all ordinary food, had subsisted for months on nothing but alcohol in one shape or another; one of these, who was bed-ridden, appeared actually fatter at the end of three months than she was at first. A surgeon's widow informed Dr. Inman, that after several successive severe illnesses, she had suffered much after her last confinement: at this

* From what has been said (in Chapter III.) it will be understood that in my opinion the tobacco materially assisted to support life.

her appetite had entirely failed her, and she had lived for many weeks on nothing but brandy and water. A surgeon at Wavertree "attended a young man with hypertrophy and patulous valves of the heart, from September 24, 1855, to April 26, 1860. For the last five years no animal food would remain on his stomach, and farinaceous he would seldom take. In the first two years brandy was the principal nutriment he subsisted on, as nothing else would remain on the stomach. Subsequently he *lived upon* this same beverage. His allowance first was six ounces of brandy, but it was gradually increased to a pint a day; he kept his flesh and good spirits nearly to the last. During the last two years he was dropsical, and he died at the age of twenty-five." Mr. Nisbet, of Egremont, communicated to Dr. Inman the case of a man in the middle class of life, who subsisted for seven months entirely on spirit and water; "he was apparently in good health and good condition." The same medical practitioner reports the case of a child affected with marasmus, who subsisted for three months on sweet whisky and water alone, and recovered; and that of another child, who lived entirely upon Scotch ale for a fortnight, and then recovered his appetite for common things. Dr. Inman himself "had a lady patient who was several times on the verge of *delirium tremens*, and he gained an intimate knowledge of her habits from personal observation, from the reports of her husband, of mutual friends occasionally residing in the house with her, of her mother, of her sisters, and of her nurse. She was about twenty-five years of age, handsome, florid and *embonpoint;* of very active habits, yet withal of a delicate constitution, being soon knocked up." This lady had two large and healthy children in succession, whom she successfully nursed. On each occasion she became much exhausted, the appetite wholly failed, and she was compelled to live solely on bitter ale and brandy and water; on this regimen she kept up her good looks, her activity, and her nursing, and went on this way for about twelve months; the nervous system was by this time thoroughly exhausted, yet there was no emaciation, nor was there entire prostration of muscular power.

The concurrence of so large a number of observations in testifying to the power of alcohol in certain circumstances to support life singly, or with entirely insignificant assistance, ought surely to produce a decided effect on the mind of any reasonable person. The want of mathematical precision, of definite evidence as to comparative weights of the body, &c., is a defect in these testimonies which must be at once conceded, but which, while it *impairs* their value, does not *destroy* it, even taking the cases singly. And after all, every possible allowance being made for exaggeration on the part of the witnesses from whom Dr. Inman, his medical friends, and myself have derived the above histories, it must in common justice be allowed that there remains a substratum which cannot be explained away, and which it would simply be dishonest to ignore. There is of course no doubt that such cases as those now repeated are rare; they are extreme and exceptional facts. But there are an immense number of other facts which approach them in character, and which any hospital physician may meet with in plenty if he looks for them. Few but those who have conducted a special inquiry in this subject, could imagine the enormous quantities of alcoholic liquors consumed by certain classes of drinkers, particularly among potmen and tavern waiters, brewers' draymen, the men employed in lading coal-barges on the river, &c. Instances are not very uncommon of men drinking, for years in succession, as much as *a pint of spirits and from half a gallon to a gallon of beer a day.** The large majority of such persons, as far as my experience goes, live upon a very inadequate quantity of solid food, even when their occupations entail constant and strenuous bodily exertion. I have been repeatedly told by such individuals, that they cannot eat, not merely because they have no appetite, but because if by making an effort they

* Many instances of equal and larger quantities of alcohol being habitually taken for years will be found in Dr. Marcet's work on Chronic-Alcoholic Intoxication.

forced themselves to take any considerable quantity of solid
food, they suffer from a feeling of heaviness and a tendency
to comatose slumber. On the other hand, there are a certain
number of intemperate drinkers whose appetite for common
food does not perceptibly fail, even while they are taking
enormous doses of alcohol; it may be that in such cases elimination takes place with unusual rapidity, and the alcohol
therefore produces little or no permanent effect upon bodily
nutrition.

The effects upon the nervous system which are usually produced in those individuals who support life chiefly by means
of alcoholic drinks are well described by Dr. Marcet, although
in his interesting tables he has not stated the particulars of
the diet of his patients. They may be summed up under the
following heads: 1. Neuralgic pains. 2. Impairment of voluntary muscular power. 3. Muscular tremor. 4. Want of sleep.
5. Hallucinations. 6. Vertigo, *muscæ volitantes*, diffused
headache. 7. Palpitation of the heart. All these are distinctly referrible to the nervous system; although Dr. Marcet,
to my surprise, has spoken of the latter as a non-nervous
symptom. Besides these, there is a long list of dyspeptic
symptoms, some of which, as *e.g.* the morning vomiting which
so commonly follows the evening debauch, appear to me
purely nervous in their origin; others depend upon the irritant action which alcohol exerts on the stomach by local contact, and are most strongly developed in persons who drink
much spirits, especially in those who drink *raw* spirits. All
the nervous symptoms are indicative of a paralyzing effect
upon the nervous system; they indicate a degree of narcosis,
but only a slight degree. The morning vomiting of the
habitual drinker, for instance, is strictly comparable to the
vomiting produced in animals (*e.g.* Experiments XV. and
XVI.) an hour or two after an injection into the peritoneum
equally as after an ingestion into the stomach.

Even these slighter narcotic effects, however, are not invariably produced by excessive habitual indulgence in alco-

hol. The physician is the last person who would desire to weaken the force of the gloomy picture which is ordinarily drawn of **the physical** debasement **of** the **excessive drinker,** but in strict truth it must be stated that there are a **few** individuals who apparently altogether escape, for **many years,** any *physical* punishment of their sensual indulgence. **The** old man whose extraordinary mode of living (almost entirely on gin) I have already described, would have been of little **service as a** practical illustration **of the bodily harm** wrought by drinking, **being, in truth, rather an usually active and** vigorous person for his time of life. On **the other hand, a** large number of topers, after continuing, **for months or years,** an utterly abnormal mode **of nutrition, develop some of the slighter degrees of chronic narcosis which depend upon** the **very gradual wearing out of the nervous system; while, in a smaller** number **of individuals, the nutrition of the nervous** centers **breaks down suddenly to such an extent as causes** *delirium tremens,* **or initiates a series of epileptic fits, or gives rise to a stroke of paralysis from the sudden giving way of** brain fibers which **have undergone white softening.**

I shall not **attempt to explain the extraordinary fact that** with such **an abnormal kind of diet as that employed by the large majority of drinkers to great excess, these accidents (and also the concurrent diseases of kidney and liver which are to be traced, when they occur, to a malnutrition of those viscera) are so long delayed, or even never occur, in certain exceptional cases. I place the fact alongside of that other, which is equally difficult to explain, of the remarkable change which so often occurs during acute disease, in cases of severe** hæmorrhage, &c., **in the behavior of the organism under large doses of alcohol. It would appear that both in the one case and the other, alcohol fails to produce its narcotic effects, and** operates merely as a **stimulant and a support capable of maintaining the chief** burden of nutrition. **Whether this is effected by means of a real** transformation of **alcohol, such as Liebig** somewhat too hastily concluded, **I am unable to say: this** much is certain—that the so-called proofs of entire elimination of alcohol in an unchanged form are as yet very inconclusive, while a great deal of evidence exists which renders such elimination improbable—that even if such proofs had been supplied, they would not disprove the possibility of alcohol acting as an aliment—and that, finally, the clinical fact is, that alcohol actually does support life in circumstances when it must, without such aid, sink from mere inanition.

I subjoin two tables which display, respectively, the operation of large doses of alcohol in nine selected cases of acute disease or injury, and the quantities of alcohol consumed by twelve topers, with its influence upon their diet; these last being also selected cases, chosen for the enormity of the excesses committed.

TABLE 1.

Disease.	Sex.	Age.	Alcoholic Liquors.	Duration of Treatment with the highest doses.	Results (as to Nervous System).
Typhoid.	Female.	24	Brandy, 12 oz. per diem increased to 36 oz. per diem.	15 days; after this greatly reduced.	No delirium after the dose had been raised; mind clear.
Flooding after labor.	Female.	32	1 bottle of brandy within 2 hours.	2 hours.	No ebriety; rather heavy sleep for 10 hours, but no true coma
Erysipelas.	Male.	43	24 oz. brandy per diem.	3 days; then reduced to 12 oz.	Delirium ceased on commencing treatment.
Erysipelas.	Male.	26	48 oz. brandy per diem.	2 days; reduced to 24, then to 12, 6, 0.	Delirium arrested; never returned.
Typhoid.	Male.	12	12 oz. brandy per diem.	4 days; half this quantity 12 days.	Delirium arrested; mind clear throughout.
Tracheotomy, for diphtheric croup.*	Female.	7	12 oz. brandy per diem.	3 days; reduced to 6, 4, 0.	No ebriety; no flushing of face.
Typhoid.	Female.	41	2 bottles port per diem.	4 days; then greatly reduced.	No ebriety; no coma, only slight drowsiness.
Pneumonia.	Male.	14 mo.	6 oz. port wine per diem.	12 days; then reduced to 3 oz.	No delirium or coma.
Bronchitis.	Female.	72	24 oz. brandy per diem.	7 days; reduced rapidly.	No delirium or coma throughout.

* My friend and colleague, Mr. C. Heath, operated: the **child made a good recovery.**
The case is reported in the *Medical Times and Gazette* for 1861.

26

TABLE 2.

	Sex.	Age.	Occupation.	Duration of Intemperate Habits.	Quantity of Alcoholic Liquors taken.	Effects upon Diet.
1	Male.	27	Tailor.	12 years.	1 pint of gin per diem and 2 pots of porter.	Eats very little solid food.
*2	Male.	83	Pensioner.	Many years.	1 bottle gin per diem for the last 20 years.	Eats one small fragment of bread in the day.
3	Male.	49	Hawker.	Many years.	About 1 pint of raw brandy per diem.	Eats no meat; only a little bread and tea.
4	Male.	29	Hawker.	10 years.	About a pint and a half of raw gin per diem.	Eats a very fair quantity of food.
5	Female.	42	None.	15 years.	About ¾ pint of brandy (with water) per diem.	Eats almost no ordinary food.
6	Male.	28	Tavern Waiter.	8 years.	4 pots of beer and 1 pint or more of spirits per diem.	Says that he hardly ever touches solid food.
7	Male.	46	Tavern Waiter.	22 years.	Has lately reached 2 pints of gin and a little beer per diem.	Eats only one small meal a day.
8	Female.	64	None.	30 years or more.	Latterly 1 pint of gin per diem.	Eats no food except biscuit; takes no tea.
9	Male.	42	Coalporter.	24 years.	For some years past 12 pints of beer a day.	Eats pretty well.
10	Male.	21	Cabman.	6 years.	For some time past 1 pint of rum per diem.	Eats little solid food.
11	Female.	21	None.	4 years.	From ½ a pint to 1½ pint of gin per diem.	Eats hardly any solid food.
12	Male.	27	Brewer's Drayman.	8 years.	2 gallons of beer per diem, a bottle of whisky every Saturday.	Eats little solid food.

* Case already described.

It must be understood that the above cases are not quoted in order to recommend a special line of practice or to guarantee its success: they stand by themselves as illustrations of a particular set of facts.

Of these twelve inveterate topers, three (Nos. 2, 10, and 12) did not suffer from consequences traceable to the alcohol at all, but came under my care for some other ailment. The remainder were all affected with some degree of chronic narcotism, and several of them had also local symptoms pointing to irritation of the stomach from frequent contact with concentrated alcoholic liquids. The severest case was in No. 5; that of a woman who had drunk secretly for years, and was a prey to the most frightful spectral hallucinations, constant muscular tremors, total want of sleep, and a strong suicidal tendency of mind. In the other cases the alcoholic symptoms were really slight, in no case making any near approach to positive *delirium tremens*, except in No. 1.

I cannot conclude these observations on the action of alcohol without noticing a novel mode in which it has been employed as a stimulant, which is likely to prove extremely important. I allude to the practice, which has recently been extensively put in force in France, of dressing surgical and other wounds, especially such as are in a sloughy condition, with alcohol. As yet I have had scarcely sufficient opportunity to test this mode of treatment, especially as surgical cases do not properly come under my hands; but there can be no difficulty, I think, in arriving at one or two positive conclusions: 1. The application of pure alcohol to the surface of a wound, whether recent, or granulating, or sloughy, forms at once a thin layer of coagulated albuminous matter, beneath the protection of which the tissues have time to regenerate themselves. 2. The subsequent applications of alcohol affect the tissues beneath the albuminous covering in a mildly stimulant manner, very small quantities only permeating, and these very gradually, as is proved by the nonoccurrence, or very rare occurrence of constitutional symptoms of alcoholism, which would otherwise certainly be produced.

THIS is the proper place for briefly noticing the action of a medicine which has come into very general use within the last few years under the name of Chloric Æther, but which is, in fact, nothing more than a mixture of rectified spirits of wine and chloroform, the latter ingredient being present in the proportion of from five or six to twelve and a half per cent. Given in very large doses, this medicine would doubtless act as a true narcotic, but in the quantities which are actually prescribed it is a pure stimulant of great value.

Of the medicinal uses of small doses of alcohol enough has already been said. The addition of small proportions of chloroform very materially heightens their effectiveness for certain purposes, particularly for subduing the slighter convulsive and spasmodic affections, and for quieting the excitement of the nervous system, in fevers and acute inflammations, and incidentally procuring natural sleep by means of this physiological action.

On the Continent, chloroform alone* (simply suspended in mucilage or some other bland fluid) has been far more employed for these purposes than in England, where the custom has so universally prevailed of giving it in alcoholic solution, that I have thought it proper to speak of the action of this medicine under the head of Alcohol, since the addition of the latter ingredient materially modifies its action, rendering it far more certain and manageable, when the mixture is properly prepared (by distillation). In doses ranging from ten minims to one drachm, "chloric æther" is a more universally serviceable stimulant, when a very powerful stimu-

---

* For illustrations of the internal use of Chloroform as a stimulant, the reader is referred to numerous cases, &c., quoted in the Bibliographical Index.

lant action is not required, than almost any other which we possess: these doses, however, apply to the standard of composition at which this medicine is fixed by the new Pharmacopœia, the "Spiritus Chloroformi" of which contains only five per cent. of chloroform.

No better instance can be given than the action of this medicine, of the efficacy of true stimulants in producing natural sleep in those exhausted conditions of the nervous system, in which, without any such grave symptoms as delirium or violent epileptiform convulsion, there is a state of restlessness and jactitation, together with a morbid wakefulness. The sleep thus induced has not in the slightest degree the character of narcotic stupor, nor is it followed by any symptoms of depression. I have repeatedly used this remedy with good effect (in the same way as I have used sulphuric æther) as a calmative in the sleeplessness of chronic alcoholism short of an actual outbreak of *delirium tremens;* and it is chiefly this kind of action of chloric æther which I would dwell upon, as its other uses are too familiarly known to require any comment or illustration. As a remedy for *pain,* chloric æther is decidedly inferior to many other stimulants, according to my experience, at least; in other respects it is an excellent representative of the class of true stimulants.

THE result of the foregoing researches may appear small as regards the amount of absolutely novel information which has been disclosed, but I venture to think that important considerations arise out of the comparison which has been instituted between the *whole* action of the three narcotics respectively. It has been too much the fashion hitherto, to compare alcohol, æther, and chloroform with each other, merely in one point of view, in that, namely, of their anæsthetic influence, and to overlook the fact that each of them possesses also stimulant properties which are in like manner strictly comparable in all three.

To speak first of the anæsthetic influence, which is common to them all when administered in large doses, it appears that the differences between their respective actions, in this direction, are more apparent than real. All of them exert a paralyzing influence upon the nervous system, which commences with the nerves of the periphery, and spreads toward the nervous centers, sensation and voluntary motion being always affected (at any rate in some slight degree) at the circumference before the general system is engaged. Thus we have feelings of numbness and loss of muscular sense, and the finer co-ordinations of voluntary motions in particular localities, before anything like general paralysis of sense or voluntary motion occurs. Thus in poisoning with alcohol or æther, we have numbness of the lips and face, and in poisoning with alcohol, æther, or chloroform, we have loss of the muscular sense of the extremities, before or coincident with the very earliest symptoms of palsy of brain or spinal cord.

Of the nervous centers the brain is the earliest to show the effects of poisoning with either of the three agents, and accordingly we get disturbance of intellect and emotions, and (almost coincidently with this) of the muscles of the eye, **which are governed by nerves** proceeding directly from the brain (motores oculi, pathetici, sixth pair).

With regard to the spinal cord, it is found that, in common with most if not all other narcotics, the three agents affect it in a certain order ; viz. that they paralyze it in successive sections from behind forward, below upward in man. There are certain special regions, however, supplied by spinal nerves, which only very slowly come so far under the influence of the anæsthetic as to lose their sensibility to pain. This was particularly observed with regard to the ano-genital region, and also with regard to the matrix of the great toe-nail; and it is interesting to observe, that all of these localities are at once singularly sensitive to pain, and singularly incapable of the distinctive sense of touch; a fact which may be verified by the use of the compass points or the anæsthesiometer. If after an operation for the evulsion of the toe-nail, at a date when the matrix has hardened but the new nail has not grown, this test be applied, it will be readily found that this structure is extremely insensitive to it. Yet the capacity of the matrix for the mere sensation of pain is so high, that a very profound degree of anæsthesia is necessary completely **to** remove it, and thus accidents have rather frequently occurred, as has also been the case with operations for phimosis, &c.

With regard to the medulla oblongata, the observed paralytic effects are confined to the later stages of narcosis, with a certain exception, namely, that the nervi hypoglossi, which **preside over** the movements of the tongue, are mostly palsied at a rather early stage, an effect which we observe most markedly in the thick and confused articulation of the drunkard, even at a comparatively early period of intoxication. The more serious effects on the medulla, such as hurried, irregular, or morbidly depressed respiration, dysphagia, &c., are produced only by a profound narcotic impression, and **are**

the precursors of death in what may be called normal cases
of fatal anæsthetic action—those, namely, in which death by
apnea takes place.

Lastly, of the paralytic affections of the sympathetic nervous
system it may be said that in the case of all these agents, the
degree in which they occur, and the parts of the sympathetic
system which they chiefly affect, have an important influence
on the course and termination of the poisoning, particularly
the degree in which the *cardiac* sympathetic is affected, since
if this be great, and rapidly produced, death by cardiac palsy
may suddenly occur.

We have, also, seen enough to satisfy us thoroughly, I
think, that neither chloroform, æther, nor alcohol, produce
the so-called "symptoms of excitement," by virtue of their
stimulant action ; but that these phenomena are, in truth, a
part of the narcotic or depressing influence upon the nervous
system. As a matter of simple **and** easily verified fact,
emotional extravagance, convulsive muscular movements,
excessive rapidity of pulse and respiration, hypersecretion
from salivary glands, from kidneys, or from skin, abnormal
formation of sugar by the liver, occur, if at all, not prior to,
but in the midst of, the paralyzing action of these anæsthetics.
So much for the induction of anæsthetic narcosis by alcohol,
æther, and chloroform. In regard to the retrocession of nar-
cotic symptoms, it may be remarked, in the first place, that
this takes place in an inverse order to that of their induction;
and, secondly, that the time occupied by it will correspond,
in the respective cases, to that consumed by the former, or
rather will be proportionate to it. Thus chloroform, which
speedily anæsthetizes, speedily leaves the organism free;
æther, which is slower in its poisonous operation, more slowly
ceases to act; and alcohol, which still more gradually nar-
cotizes, continues its effects much longer than either. And
this brings us to consider the question of elimination, as it
applies to the three anæsthetics comparatively.

It is obvious that the rapidity with which the three anæsthetics are respectively eliminated is in inverse ratio to their solubility in the blood. Chloroform, which is soluble only in minute proportions, is eliminated altogether, or nearly so, by the lungs, within an hour or two. Æther, which is to a considerable degree soluble in serum, appears to be eliminated to a very small extent by the kidneys; but is to be detected in the breath for many hours after an anæsthetic dose has been administered, though I have doubts whether this is the case when merely medicinal doses have been used internally, in cases of disease to which its action is beneficial. But alcohol, which is highly soluble in serum, is eliminated only in minute proportions, and very slowly, except in instances where an excessive dose, and such as produces narcotic effects, has been suddenly thrown into the system. Even in the latter case, it is with difficulty that the organism is so far freed of its presence as to emerge from the narcotic state.

I have dwelt on this fact with regard to alcohol, and have detailed experiments which exemplify it, because I am anxious to impress on the reader's mind an important inference which appears to follow inevitably from it: namely, that alcohol was never designed by the wisdom of Providence to be employed by the human race *as an anæsthetic at all*, but for the sake of those stimulant qualities of its non-narcotic doses, which are to a certain extent also shared by small doses of æther and chloroform. It seems as if the former were intended to be the medicine of those ailments which are engendered of the *necessary* every-day evils of civilized life, and has therefore been made attractive to the senses, and easily retained in the tissues, and in various ways approving itself to our judgment as *a food;* while the others, which are more rarely needed for their stimulant properties, and are chiefly valuable for their beneficent temporary poisonous action, by the help of which painful operations are sustained with impunity, are in great measure deprived of these attractions, and of their facilities for entering and remaining in the system.

It is most interesting, further, as supporting this view, to observe the difference which exists between alcohol on the one hand, and æther and chloroform on the other, as to the facility with which the nervous system becomes accustomed to their use in narcotic doses. Every one knows (although it is not true as stated by teetotallers that the use of *moderate* quantities of alcohol becomes after a time ineffective, and the dose must be increased) that there *is* a fatal necessity **for the** individual who has once habituated himself to the *narcotic* effects of alcohol to go on augmenting his daily allowance. With regard to æther, however, this is hardly at all the case, and with chloroform still **less. We** do, indeed, hear of wonderful instances of persons becoming able by practice to take large doses of æther or chloroform without any narcotic effect being produced ; but I unhesitatingly state that these reports are based on a misconception. I have repeatedly had to administer chloroform daily for weeks or even months together to the same patient, and have never found this necessity for increasing the dose; and I make no doubt that the mistaken idea has originated in the wasteful and dangerous practice of entrusting the administration to an unskilled hand, and of employing a mere handkerchief, or some other unscientific apparatus, with which it is literally impossible to form even an approximative idea of the proportion of the anæsthetic which really enters the blood. The simple fact is, not that the patients become able to inhale large doses, but they get progressively more careless and wasteful, and accordingly allow more and more of the chloroform or æther to escape into the air of the room. All statements as to "tolerance" which have been based on such administrations as these are utterly worthless.

Moreover, with regard at least to chloroform, there is an almost utter absence of that harmlessness of large doses in **acute** diseases, involving the nervous system, which is so remarkable a peculiarity of alcohol. Even in tetanus, when, **if ever**, we might expect that chloroform might be used recklessly with impunity, there is no abnormal tolerance for chloroform whatever. I speak with justifiable confidence on this point, because I have had large experience : and I affirm that full narcosis may be produced with precisely the same doses in tetanus as in health, and that it is just as easy to produce serious and even fatal results by the rapid saturation of the blood with a large dose as in health.

There is a real difference, we see, between the limits of the stimulant action of alcohol on the one side, and of that of æther and chloroform on the other. It is, therefore, the less surprising to find that the former possesses as decided a superiority in regard to its power to sustain vitality, particularly in the absence of a sufficient supply of ordinary food. It is not that æther and chloroform may not act in the same sense as foods, for though they both, the latter especially, are dismissed from the organism with much greater rapidity than alcohol, there can be no doubt that minute doses of them really do restore nervous vigor in certain circumstances, in a way which we do not in the least explain by the mere statement that they have acted as stimulants. But they will not, like alcohol, produce these effects in the prolonged absence of common nutriment, and they cannot be regarded as substitutes for alcohol in this respect.

I shall conclude with a few remarks on the reasons which have induced me to refrain from discussing, at present, the evidence which exists as to the influence of alcohol on the elimination of matters representing the *waste of the tissues*.

I consider that it is at present impossible to discuss this question with good effect, because we do not know the conditions in which our experiments would be decisive. With every respect for the untiring energy and perseverance which have distinguished the researches of Dr. E. Smith, of MM. Böcker, Vierordt, Lehmann, and others, I cannot think that the results they afford are such as can settle the question of the alimentary value of a substance like alcohol, because in truth we are not as yet in possession of any certain knowledge as to the physiological significance of variations in the amount of some of the most important elements of the bodily excretions. If the reader will reflect on the extreme complexity of the problem, on the extraordinary fluctuations of opinion which have taken place within a few years on these subjects, if he will recall to mind the numerous instances in which the most **elaborate** experiments have proved to be almost valueless, he **will** easily understand why, for the present, I judge it more useful to deal with simpler facts. Upon such a question as that of the exact meaning of a given increase in the carbonic acid expired during any particular period of twenty-four hours, there is room for endless difference of opinion, in the present state of our knowledge. I have never been able, for instance, to perceive the logical necessity of the inference that, because a large excretion of carbonic acid in the breath coincides with the digestion of certain undoubted foods, the alimentary value of the latter was to be considered as directly proportionate to such increase. Carbonic acid appears, as far as we can judge from late researches, to represent, much more nearly than any other element of excretion, the amount of *organic commotion*, if I may use such a phrase, which is taking place. But it has been my own object throughout this work to enforce the truth that commotion—or "action," as it is often loosely called—is not the measure of vital power: that the latter is the result of a special balance and proportion of matter and forces which is characteristic of the "individual," or "living creature:" that consequently, unless we know all the conditions of the problem of life, we are not entitled to say that increased development of a particular force implies anything more than derangement of the balance, and that after all it were far safer to rely on the teaching of experience at the bedside, and of the daily practice of large classes of **men** whose dietetic habits the physician necessarily becomes familiar with, than on the dicta of a science like physiological chemistry, which, notwithstanding its rapid progress of late years, is still in a merely rudimentary condition.

NOTE A.

BIBLIOGRAPHY OF THE DOCTRINE OF VITAL SPIRITS.

It has occurred to me, that to any one who desires to satisfy himself as to the origin and history of this doctrine, a more copious list of references to the sources of information, especially as regards early authors, may be not unwelcome, as saving much preliminary labor. In the following bibliographical table, added to the notes appended to the text of Chap. I, the reader will at least find sufficient of material to start with: above all, in the numerous references to Galen now given, he will find the means of arriving at a tolerably clear idea of the way in which the *medical* doctrine of vital spirits (as distinguished from the ecclesiastical theory which grew out of special interpretations of certain phrases in Holy Scripture) grew up. It was evidently by Galen's instrumentality that the Platonic theory of the ψυχῆς θνητὸν γένος received consistency and form, and was made the means of explaining the facts of life and disease, of which he was so acute an observer.

Aristotle, De Animâ, I, 2; II, 4, 8; III, 7. Hist. Animal, I, 16.
Dandini. Comment. in Aristot. de Animâ.
Barthélemy Saint-Hilaire. Psychologie d'Aristote.
Comment. Conimb. in Aristot. de Animâ.

Justin Martyr (Migne's edition). See also Tennemann's Geschichte der Philosophie, 11 vols. Leipzig, 1819.
Tertullian, De Animâ, especially cap. X, XI, and XV (soul corporeal, nourished by the inspired air).
Lactantius, De Opificio Dei, cap. XI (soul nourished by external air).
Nemesius, De Naturâ Hominis, cap. 2, 4 (Ed. Matthiæ).
Athanasius, De commune essentiâ Patris et Filii et Sp. Sancti. Opera Omnia, vol. II, p. 239. Paris, 1627.
Augustine (Caillau's Patres ecclesiæ sæculi quinti), De animâ et ejus origine (in lib. X, of De Genesi ad litteram; here he argues against Tertullian's doctrine of the corporeity of the soul), De quantitate animæ, cap. XXIII, XXX, XXXIII (argues against the Aristotelian doctrine of the *anima per totum corpus diffusa*).
Irenæus, I, 5.

Galen, De usu part. VI, 17; De utilitat. resp. 5; De usu part. VII, 8; De Hipp. et Plat. dec. VII, 13; De Hipp. et Plat. dec. III, 8; De usu part. IX, 4; De Methodo Medendi, XII, 5; De usu pulsuum, cap. 2; De Instrum. Odor. cap. 3; De usu part. VI, 13; In prognost. Hippoc. I, 23; De fœt. format. cap. 6; De difficult. resp. II, 7; De utilitat. resp. 5; De Hipp. et Plat. dec. VII, 3, 4. (The above references are to the Junta Latin edition published at Basle, 4 vols., folio, 1560.)

Plotinus, Enneades, I, ɪ, 7; IV, ɪɪɪ, 19, 23, 29; IV, vɪɪ, 4; II, ɪ, 1, 7; V, ɪ, 6, 7.

Porphyry. Vita Plotini.

Steinhardt, Meletemata Plotiniana.

Apuleius, De vit. Philosoph. II; De dogm. Plat. p. 51.

Plutarch, Physic. philos. decret. IV, 5, 84; V, 24, 124.

Virgil, Æneid VI.

Epicurus, in Diogen. Laert. lib. X, s. 66, p. 630.

Asclepiades, in Sext. Empiric. lib. II, s. 7, p. 460.

Aretæus, Caus. acut. lib. II, c. 1.

Cassii Iatrosophistæ quæstiones natur. et medicin. Ed. Gessner. Venet., 1562.

Garbo, Expositio super capit. de generatione, f. 20 b.

Thomas Aquinas, De summâ theologiâ, pars prima, quæst. 76, art. 3; quæst. 75. art. 3; quæst. 78, art. 2 (*principium corporis formativum*); quæst. 119, art 1 (*humidum radicale*, and *humidum nutrimentale*); pars secunda, quæst. 38, art. 5.(The *principium corporis formativum* is extremely like the σπερματικὸς λόγος of Plotinus.)

Gilbertus Anglicus, Compendium Medicinæ, f. 118 b; 242 a, 245 b.

Torrigiano, or Turrisanus, lib. I, f. 11 a.

Theophrastus, De Modo pharmacandi, lib. II, p. 772.

Paracelsus, Magna Philosophia. (Ed. Dordr.) p. 176; Paramirum, II, p. 36; IV, p. 77.

Robertus Fludd, M.D. (a Rosicrucian), Utriuspue cosmi . . . . metaphysica physica et technica historia. Francofurti, 1617.

Robertus Fludd, M.D. Answer to M. Foster, &c. (being a defense of the weapon salve).

Marsilius Ficinus, Opera, in 2 tomos digesta. Basle, 1561.

Marsilius Ficinus, De vitâ triplici (4to. Bonon, 1501), cap. 2.

Gassendi, Syntagma Philosophi Epicuri. 4to. London, 1668.

Charleton, Physiologia Epicureo—Gassendo—Charletoniana, &c. London, 1654.

Francis Glisson, Tractatus de Naturâ substantiæ energeticâ, seu de vitâ naturæ, &c. London, 1672. 4to.

Thomas Willis, De animâ brutorum, &c. 8vo. London, 1672.

Barthez, De princip. vit. hominis. 1772.

Stahl, De verâ Diversitatite corporis mixti et vivi. 4to. Hallæ, 1707. Theoria Medica vera. 4to. Hallæ, 1708.

Hermann Boerhaave, Praxis Medica, &c. 5 tom. 8vo. Petav., 1728.

Fred. Hoffmann, Commentarius de differentiâ inter ejus doctrinam medico-mechanicam et G. E. Stahlii doctrinam medico-organicam.

NOTE B.

ON THE PHYSIOLOGY OF THE PUPIL-CHANGES IN NARCOSIS.

As the opinions expressed on this subject (in Chap. V) are opposed to those entertained by physiologists in general, I feel bound to offer some explanation of my reasons for inclining to those opinions.

The controversy as to the nature of the machinery concerned in the movements of the iris is a very old one, at least as old as the days of Haller. The point chiefly at issue has of course been this—whether the changes in the diameter of the pupil were due simply to muscular contraction, or to the alterations in an erectile tissue?

On the part of the muscular theory, it has been urged that there are plain indications of the existence of two muscles, a dilator and constrictor, the former consisting of fibers radiating from the pupil to the circumference of the iris, the other consisting of circular fibers arranged round the margin of the pupil. It is known, from the researches of Valentin, Budge, &c., that there are two sources of nervous influence on the iris: viz. the ciliary branches of the motor oculi, and the sympathetic branches from the ophthalmic ganglion (which represents the junction of the fifth nerve with the cranial sympathetic). The former nerves, then, would be the channel of nervous influence from the brain, the latter from the cilio-spinal region of the cord. This anatomical basis being presupposed, we may see how the physiology of the iridal movements caused by poisoning would be explained, on the principles currently received, by referring to a recent work of a most able physiologist, which incidentally treats of this matter.[*] Dr. Fraser remarks that "a little consideration will show that the pupil may be influenced in at least six different methods. 1. By cerebral irritation. 2. By cerebral depression. 3. By spinal irritation. 4. By spinal depression. 5. By a combination of cerebral irritation with spinal depression. 6. By a combination of cerebral depression with spinal irritation." He gives certain examples of the different toxic actions; first, "two *cerebral*" groups, the *one* causing irritation of the brain and contraction of the pupil (*e.g.* opium), the other causing depression of the brain and dilatation of the pupil (*e.g.* Belladonna, Æthusa Cynapium, Hyoscyamus, Alcohol, White Hellebore). Next, "two *spinal*" groups, the one causing spinal irritation, and dilatation of the pupil (*e.g.* Strychnia), the other spinal depression and contraction of the pupil (*e.g.* Calabar Bean, Aconite). And, finally, "two combined cerebral and spinal" groups, the one causing cerebral irritation and spinal depression, and consequent contraction of the pupil (*e.g.* Rue), the other causing cerebral depression and spinal irritation, and consequent dilatation of the pupil (*e.g.* Cicuta virosa, Nicotine, Prussic Acid, Digitalis).

These minute and complicated theories are probably required in order to account for the phenomena of the pupil-changes caused by the different narcotics, *if* we accept the muscular theory of iridal movement. They are, however, quite inconsistent with the universal and inevitable sequence of narcotic phenomena, as experimentally observed on the large scale. It is evident that *all* the phenomena of narcosis, whatever their external appearance, result from devitalization of that part of the nervous system which is immediately concerned in their production; the idea of vital exaltation in the

[*] On the Calabar Bean. By Thomas R. Fraser, M.D., &c., &c. Edinburgh: Oliver and Boyd. 1863.

brain, for instance, going on simultaneously with vital depression in the spinal cord (or *vice versâ*) is inconsistent with what any unprejudiced experimenter must observe, viz. that the process of narcotism is an *uniform* one, and tends entirely in the direction of nervous death.

On the other hand, the *erectile* theory of iridal movements has received a most important reinforcement within the last few years. Not to dwell on what has been advanced by Dr. Fleming,* in connection with the physiology of poisoning with Atropia, I may refer to the masterly exposition of the anatomy of the iris, given by M. Rouget, in a paper read by him before the Société de la Biologie. The subsequent discussion of M. Rouget's views, in a hostile sense, by M. Sappey, appears to me to heighten their value, since even the latter most able anatomist could produce no arguments which seriously affected their validity.

The chief points established by M. Rouget are these: 1. The continuity of the vascular elements of the iris with those of the choroid. 2. The continuity of the muscular fibers of the iris with those of the ciliary muscle. 3. The discovery that the superficial appearance of radiation in the iris by no means really indicates the course of the muscular fibers, but merely that of certain folds of the vascular tissue, which are, in truth, a continuation of the ciliary process. These folds can be seen to be thrown into zigzags during dilatation, under a sufficiently powerful microscope. 4. The impossibility, except by the use of high powers, of microscopically demonstrating the true course of the fibers-proper of the iris: they are arranged, not in a regularly radiating direction, but obliquely, so that they intersect at various points in their progress from the pupil to the corneal edge.

With regard to the circular fibers supposed to surround the pupil, the existence of these in man has been repeatedly denied by the most competent anatomists; and in our own times by Robin and Hyrtl. It is scarcely too much to say that, but for the exigencies of the muscular theory, they would never have been seriously put forward as assisting in any material degree the contractions of the human pupil.

On the whole, joining this latter fact with the brilliant researches of M. Rouget, I cannot but feel that the evidence greatly preponderates in favor of the view that the iris is an erectile tissue—an offshoot of the choroid tunic—and that its muscular elements subserve merely the same office as that of the like elements in erectile structures elsewhere. Adopting this opinion, the following appears to me the natural explanation of the changes in the pupil produced by various narcotic poisons:

When a narcotic poison operates with *medium* force upon the cranial sympathetic system, contraction of the pupil will be one of the symptoms produced. The effect will be equivalent to that produced when the cervical sympathetic cord is divided; such a procedure by no means cuts off all nervous influence upon the iridal vessels; the vitality of the nerve beyond the section, at least, will remain in great part unimpaired, to say nothing of possible communications with the centers by circuitous routes. Under such circumstances, the condition of the iridal vessels is by no means one of inactivity; on the contrary, the blood circulates through them with increased force, at the same time that their caliber is enlarged. Contraction of the pupil results.

But, in the case of *extreme* narcosis of the cranial sympathetic, things are otherwise. Here the sympathetic branches may well be supposed to be, at least, temporarily dead, down, to their ultimate filaments. Moreover, the muscular coats of the vessels have

* Dr. Fleming's paper will be found in the *Edinburgh Monthly Journal* for 1863.

themselves, in all probability, been deprived of their vital qualities; the *rigor mortis* of arteries sets in (a thing which often happens locally, before general death of the organism), and the blood is expelled from the iridal arteries, capillaries, and veins. Dilatation of the pupil results.

Evidence might be accumulated in favor of this view to a large extent; but I prefer to give one or two illustrations of the advantages it possesses over current theories, by considering the case of one or two poisons of marked and unmistakable action.

What, for instance, is the effect on the pupil of acute prussic acid poisoning? Dilatation, usually to an extreme degree. Dr. Fraser explains this result by declaring, that prussic acid irritates the cord and depresses the brain, and thus simultaneously paralyzes the constrictor, and excites the dilator, fibers of the iris. On the contrary, I maintain that prussic acid kills the spinal cord; and that it operates with especial violence on that part of the cord known as the cilio-spinal region, I consider proved by the flushed face and protruberant glittering eyes of persons poisoned with it. And, on the principle already explained, I submit that the progressive devitalization of the pupillary nerves which would naturally result, and the ultimate molecular death of the blood-vessels which would quickly follow (owing to the rapid deoxygenation of the blood which prussic acid causes), will sufficiently account for the dilatation of the pupil, without involving a theory so improbable on the face of it as that which assumes the simultaneous occurrence of an increase in spinal, and a diminution of cerebral, vitality.

Again, Dr. Fraser speaks of opium as a poison that contracts the pupil; and he explains its action by asserting that it irritates the brain. But opium, in its *final* effect even on the human being, and much earlier in some lower animals (*e.g.* the cat), usually produces *dilatation* of the pupil. In truth, I believe, it is a mere question of the cranial sympathetic being the *ultimum* or the *primum moriens* that determines the perseverance of contraction up to the moment of death, or its yielding to dilatation before that event occurs.

Finally, I may refer to the action of the Calabar bean as affording, in my opinion, an instance of the inaccuracy of the ordinary statements about the causes of the pupil-changes in narcosis. The Calabar bean has been represented as having an *essential* action on the iris, in virtue of which it causes the pupil to contract. But it does not produce this effect when the blood is fully poisoned. It is only the slighter influence, which can alone be produced on the iridal movements by local applications, which causes contraction. A dose of the bean (taken internally) such as is sufficient to produce acute general poisoning, invariably falls with destructive force upon the sympathetic system, producing paralysis of the heart, and *dilatation* of the pupil.

The above considerations, and others which might be adduced, have compelled me to believe that, at least in narcosis, the pupil-changes are not primarily caused by contractions of any special muscles, but by vascular alterations. With regard to the interesting researches of Brown-Séquard, on the contraction of the pupil after death (*Journal de Physiologie,* 1859), which might at first be supposed to be fatal to this theory, three things may be remarked: 1. That the experiments urgently need to be repeated a great many times, as, from the great difficulty of correctly observing the facts, there is infinite room for mistake; 2. That the contractive changes observed are unlike those in narcosis, in many respects; 3. That even if the observations be perfectly accurate, they admit of a possible interpretation different from that given by their author, viz. that the action of light in some way restores, in a measure, the vital

27

status of the muscular fibers of the iridal vessels, and also of radiating fibers of the iris described by Rouget; and thus puts an end to the condition of rigor mortis in which they had previously been.

NOTE C.

ON THE FOOD ACTION OF COCA.

Interesting information on this subject is to be found in the prize treatise of Dr. Mantegazza, published at Milan, 1859, and which I had not seen at the time the remarks on this subject in the text were written. The general conclusions at which the author arrives are the following: 1. The leaves of coca, chewed or taken in a weak infusion, have a stimulating effect upon the nerves of the stomach, and facilitate digestion very much; 2. In a large dose, coca increases the animal heat, and augments the frequency of the pulse, and consequently of respiration; 3. In a medium dose, three to four drachms, it excites the nervous system in such a manner that the movements of the muscles are made with greater ease—then it produces a calming effect; 4. Used in larger doses it causes delirium, hallucinations, and congestion (?) of the brain. (This last statement rests upon insufficient evidence, and ought not to be received.) Dr. Mantegazza is quite prepared, from his own personal experience of the singular effects produced by coca-chewing, to credit the accounts given by so many South American travelers as to the power of coca to sustain life for a considerable period in the absence of any other nourishment, and that the Indian mail-carriers *often travel for three or four days without any food except coca.* An abstract of Dr. Mantegazza's work may be found in the *Pharmaceutical Journal* for 1860 (Vol. I, Second series).

In Bouchardat's "Annuaire Thérapeutique" for 1864, is an abstract of some valuable observations of Dr. Reis, on the action of coca. According to Dr. Reis, a very considerable dose is required in order to produce any quickening of the heart's action, and it is possible to produce all the stimulating and enlivening effects of the drug without any of the quasi-febrile excitement which characterizes the action of large doses. The account given by Dr. Reis is very valuable, because it illustrates in a most forcible way that radical difference between large and small doses of narcotic-stimulants which I have endeavored to impress on the reader. The lesson inculcated is of the more weight as the author is quite free from any bias toward the **views** supported in this work, and persists, on the one hand, in speaking of the poisonous effects of large doses of coca as "stimulation," and, on the other, in denying that coca can have any true food action.

NOTE D.

BIBLIOGRAPHY OF THE ACTION OF ALCOHOL, CHLOROFORM, AND ÆTHER.

The following bibliographical index by no means pretends to be a complete table of the literature of the subject. It is merely inserted here in order that the reader may not suffer from the want of distinct references to the sources of information which the foregoing "Researches" have more or less directly occasioned. It was clearly impossible for me to find space for the statement of my own investigations, limited though they be, except by avoiding all unnecessary crowding of my pages with discussions and with foot-notes.

Aran. Anæsthesia produced by Chloroform taken into the Stomach. Union Médicale, 1852, t. VI, p. 37.

Alienza, Ramon. Innerlich Anwendung des Chloroform bis Wechselfieber. Schmidt's Jahrbuch, CXI, 284.

Böcker, F. W. Wirkung des Bieres auf den Menschen. Schmidt's Jahrbuch, LXXXIII, 165.

Baudot, Edmond. De la destruction de l'alcool dans l'organisme. Union Medicale, Novembre et Décembre, 1863 (four papers).

Boucard, F. De la nature de l'operation anæsthésique de l'éther et du chloroforme. Gaz. des Hopitaux, 13, 1856.

Broxholme, F. G. Hernicrania Quotidiana cured by Chloroform. Lancet, June, 1849.

Beyran. Observations de syncope provoquée par l'inhalation du chloroforme ayant duré une heure et demie. Bull. de Therap. Mars, 1852.

Black, C., M.D. How shall we ensure Safety in the Administration of Chloroform? London, 1855.

Biswanger (Martin and). Das Chloroform in seinen Wirkungen auf Menschen und Thiere. Leipzig, 1855.

Berchon, E. De l'emploi des anæsthésiques, &c. Paris, 1861.

Bocanu. Règles pour l'administration du chloroforme. Gaz. des Hop. 41, 1852.

Bouisson. Parallèle de l'éther sulf. et du chloroforme. Gaz. Med. Fev., 1849 (two papers).

Batailbé (et Guillet). De l'alcool et des composés alcooliques en chirurgie. Pamphlet 8vo. Paris, 1859.

Bouchardat (Annuaire de Thérapeutique, 1844). Internal use of Liquid Chloroform in Asthma.

Bernard, Claude. "Leçons sur les effets," &c. (quoted in the notes to the text), and numerous other works of this author deal with the comparative action of alcohol, chloroform, and æther.

Béclard, J. Traité élementaire de le physiologie humaine (op. cit.).

Blot. De l'anæsthesie appliquée à l'art des accouchements. Paris, 1857.

Bouchardat et Sandras. De la digestion des boissons alcooliques, &c. Ann. de chimie et de phys. 1847.

Bouchardat. Action comparée des boissons alcooliques sur les animaux. Nouvelle Encyclop. des Sciences Médicales. 1846.

Brodie, Sir Benjamin. Physiological Researches. London, 1851.

Beau. Études analytiques de physiologie et de pathologie sur l'appareil spléno-hépatique. Arch. Gén. de Médecine, 1851. 4to., serie t. XXV, p. 406, et seq.

Blandin. Amputation during the Anæsthesia of Drunkenness. Bull. de l'Academie, XII, 317.

Coates, Martin. The safe Administration of Chloroform.

Chambert. Des effets physiologiques et thérapeutiques des éthers. 1848.

Coze. Tetanic Contraction of Muscles produced by Injection of Chloroform into Arteries.

Comptes Rendus de l'Academie des Sciences, 1849, t. XXVIII, p. 531.

Christison, Dr. On Poisons. 3d edit. Edinburgh, 1845.

Carpenter, Dr. W. B. The Physiology of Temperance and Total Abstinence. London, 1858.

Carpenter, Dr. W. B. On the Use and Abuse of Alcoholic Liquors. London, 1850.

Cohn. Ueber die tödtlichen Causal momente bei Pneumonia-potatorum deren Diagnose und Verwirthung für die Therapie. Schmidt's Jahrbuch, LXXXIX, 197.

Calmeil. De la paralysie générale chez les aliénés. Paris, 1826.

Clendon, J. C. On the Use of Chloroform in Dental Surgery. London, 1849.

Duchek. Ueber das Verhalten des Alkohols im thierischen Organismus. Prag. Vierteljahrsch. X, 3, 1853.

Dagonet. Chloroforme. Gazette Médicale de Strasbourg, 4, 1852.

Duroy. Procédés pour découvrir le chloroforme dans le sang et dans les cadavres. Journ. de Pharmacie, Avril, 1851.

Duméril and Démarquay. Recherches expérimentales sur les modifications, &c. Paris, 1848.

Ford, W. Hudson. On the Normal Presence of Alcohol in the Blood. Schmidt's Jahrb. CXII, 148.

Fourget. Action de Chloroforme. Gaz. Méd. de Strasbourg, 4, 1852.

Flourens. Recherches expérimentales sur les propriétés et les fonctions du système nerveux, &c. 2d edit.

Faure. Chloroforme et asphyxie. Arch. Gén. de Médecine, 1858, t. XII, p. 170.

Gorup-Besanez. Ueber den Einfluss des Ætherismus auf die Blutmischung. Schmidt's Jahrb. LXV, 8.

Guérin. Injection of Chloroform into the Veins. Bull. de l'Académie, 1848, t. XIV, p. 297.

Gosselin. Recherches sur la morte subite par l'influence du chloroforme. Arch. Gén. t. XVIII, p. 387.

Glover. On Bromine and its Compounds. Ed. Med. and Surg. Journ., vol. 58.

Guyot. De l'emploi de l'alcool comme méthode abortive des fièvres d'accès. Union Médicale, 1860, 108.

Huss, Magnus. Chronische Alkohols-Krankheit (aus dem Schwedischen übersetzt von Gerh. van den Busch), 1852. Leipzig and Stockholm.

Hammond, Dr. W. A. The Physiological Action of Alcohol and Tobacco upon the Human Organism. American Journal of Medical Sciences, October, 1856.

Harley, Dr. G. Nouvelle méthode pour produire artificiellement la diabète. Note lu à la Société de Biologie de Paris, 1853.

Harley, Dr. G. The Chemistry of Respiration. Brit. and Foreign Med.-Chir. Review, October, 1856.

Harley, Dr. G. The Physiology of Saccharine Urine. Brit. and For. Med.-Chir. Rev., July, 1857.

Herard. De l'utilité des boissons alcooliques au debut des accès de fièvre intermittent. Gaz. des Hop. 88, 1861.

Jacobi. Experimentelle Untersuchungen über die Wirkungen des Alkohol mit besonderer Rücksicht auf die Grade der Verdunnung mit Wasser. Deutsche Klinik, 23, 26, 31, 34. 1857.

Ivonneau (translated by Hartmann). Das Chloroform und seine Anwendung in der äussern und innern Heilkunde. Weimar, 1854.

Klencke. Untersuchungen über die Wirkungen des Branntweinsgenusses auf den lebenden Orgauismus.

Lehmann. Précis de physiologie expérimentale.

Longet Expériences rélatives aux effets de l'inhalation de l'éther. Paris, 1847.

Lallemand, Duroy and Perrin. Du rôle de l'alcool et des anæsthésiques. Paris, 1860.

Lach. Des causes de la mort par chloroforme. Gaz. Méd. de Strasburg, 2, 1852.

Liebig. Animal Chemistry in its Relatiou to Physiology and Pathology. Third edition (translated by Dr. Gregory). London, 1846.

Lassaigue. Experiments on the Mixture of Æther Vapor with Atmospheric Air. Bull. de Acad. t. XII, p. 446.

Macnish. The Anatomy of Drunkenness. Glasgow, 1828.

Marcet. Ou Chronic Alcoholic Intoxication. 2d edit. London, 1863.

Masing. Ueber die Veränderungen welche mit genossenem Weingeist im thierischen Körper vorgehen. Inaug. dissert. Dorpat, 1854.

Murphy, Dr. Chloroform in the Practice of Midwifery. London, 1848.

Murphy, Dr. Further Observatious on Chloroform in the Practice of Midwifery. London, 1850.

Magendie. Precis élementaire de physiologie. 4th edition.

Magendie. Phenomenes physiques de la vie.

Orfila. Traité du Toxicologie. 4th edition.

Ogston, Dr. F. Phenomena of the more advanced stages of Alcoholic Intoxication. Edin. Med. and Surg Journal, Vol. 40, p. 276.

Prié. Considerations sur la digestion des substances animales. Paris, 1837.

Percy, Dr. J. An Experimental Inquiry iato the Presence of Alcohol in the Ventricles of the Brain. London, 1859.

Perrin. Reply to M. Baudot's papers on Alcohol. Union Médicale, 24 Décembre, 1863.

Pavy, Dr. On the Nature and Treatment of Diabetes. London, 1862.

Rayer. Mémoire sur le delirium tremens. 1819.

Rayer. Injectious of Alcohol into the Peritoneum of Animals. Dict. de Méd. et de Chir. I, p. 291.

Royer-Collard. De l'usage et de l'abus des boissons fermentés, &c. Paris, 1838.

Reynoso. Artificial Diabetes. Comptes Rendus, tom. XXXII, p. 416.

Roësch. Dè l'abus des boissons spiritueuses. Ann. d'Hygiène, 1838, t. XX, p. 246.

Ramdohr. Leber und Gehirn bei chronischer Alkohol-intoxikation. Schmidt's Jahrbuch, XCVIII, 6.

Renault. Experiments on the Dosage of Æther in the production of Anæsthesia. Bull. de l'Acad. 1848, t. XIV, p. 281.

Ragsky. Detection of Chloroform in Blood. Journ. de Pharmacie. Août, 1850.

Robert. Des règles à servir dans l'administration des anæsthésiques. Paris, 1859.

Schmidt's Jahrbuch. Additional cases of Chloroform Treatment in Intermittent Fever. Vol. LXIX, 57; LXXVII, 73.

Sichel. Amaurosis from Alcoholic Delirium. De l'amaurose, p. 713.

Smith, Dr. Edward. Ou the Action of Alcohols. Loudon, 1861. (Pamphlet.)

Smith, Dr. Edward. On the Action of Foods upon the Respiration during the primary processes of Digestion. Transactions of the Royal Society, 1859.

Smith, Dr. Edward. On the Action of Alcohols in the Treatment of Acute Diseases. Transactions of the Medical Society of London, 1861.

Sédillot. De l'insensibilité produite par le chloroforme et l'éther. Paris, 1848.

Scoutetten. Recherches sur les anæsthésiques en général. Metz, 1858.

Simpson, Dr. J. Y. Account of a New Anæsthetic Agent as a substitute for Sulphuric Ether in Surgery and Midwifery. Edinburgh, 1847. (Pamphlet.)

Simpson, Dr. J. Y. Remarks on the superinduction of Anæsthesia in natural and morbid Parturition. Edinburgh, 1859. (Pamphlet.)

Simpson, Dr. J. Y. Eclampsia neonatorum, and Chloroform. Edin. Monthly, January, 1852.

Société des Médecins Allemands de Paris. Recherches et expériences sur l'inhalation de l'éther sulfurique. Gaz. Médicale, 6 Fevrier, 1847.

Snow, Dr. J. On Chloroform and other Anæsthetics. London, 1858.

Snow, Dr. J. Narcotism by Inhalation of Vapors (reprinted from Medical Gazette). London, 1848.

Serres. Local Application of Æther to Nerves. Comptes Rendus de l'Acad. des Sciences.

Société Méd. d'emulation. Recherches sur les moyens à employer contre les accidents, &c., t. XXIV, pp. 162 and 227. Union Méd. t. IX, p. 33.

Sutton. Tracts on Delirium Tremens, &c. London, 1813.

Schiff. Untersuchungen über die Zuckerbildung in der Leber. Würzburg. 1859.

Sancery, Dr. (Use of æther in acute mania) Betrachtungen über Ergebnisse im Irrenhause von Fains, 1848.

Saurel. Use of Anæsthetic Inhalations in Internal Diseases. Gazette Médicale, 6, 7, 11, 12 (1851).

Spengler. Chloroform-Inhalation bei innern Krankheiten. Schmidt's Jahrbuch, LXXXIX, 183.

Symonds, Dr. On Death by Chloroform. Med. Times and Gazette, March, 1856.

Trotter. An Essay, &c. on Drunkenness, and its effects on the human body. London, 1804.

Thomeuf. Essai clinique sur l'alcoolisme. Paris, 1859.

Tardieu. Observations médico-légales sur l'état d'ivresse (Annales d'Hygiéne, 1848, t. XL, p. 390).

Tourdes. On the Phenomena of Death from Ether and Chloroform. Communicated to the Société de Médecine de Strasbourg. 1847.

Townley, Dr. Parturition without pain. London. 1863.

Thompson, W. H. Pathology of Tetanus (treatment with diffusible stimulants). American Medical Times, February, 1861, Nos. 6 and 7.

Ville et Blandin. Researches on Influence of Etherization on the Exhalation of Carbonic Acid. Comptes Rendus, 1847, XXIV, 1017.

Vierordt. Physiologie des Athem. Karlsruhe, 1845.

Wohler. On the Non-elimination of Alcohol by the Kidneys. Journal des Progrès, XI, p. 109.

Weber. Chirurgische Erfahrungen, &c.